CHEMISTRY LIBRARY

APR 2 3 1975

UNIVERSITY OF MARYLAND
COLLEGE PARK, MARYLAND

New Developments in

TITRIMETRY

TREATISE ON TITRIMETRY

Editor in Chief
JOSEPH JORDAN
Department of Chemistry
The Pennsylvania State University
University Park, Pennsylvania

VOLUME 1: Inorganic Titrimetric Analysis: Contemporary Methods
  Walter Wagner and Clarence J. Hull

VOLUME 2: New Developments in Titrimetry
  edited by Joseph Jordan

OTHER VOLUMES IN PREPARATION

New Developments in TITRIMETRY

Edited by Joseph Jordan

Department of Chemistry
The Pennsylvania State University
University Park, Pennsylvania

MARCEL DEKKER, INC., New York   1974

Chem.
QD
111
.J6

COPYRIGHT © 1974 by MARCEL DEKKER, INC. ALL RIGHTS RESERVED

Neither this book nor any part may be reproduced or transmitted in any form or by any means, electronic or mechanical, including photocopying, microfilming, and recording, or by any information storage and retrieval system, without permission in writing from the publisher.

MARCEL DEKKER, INC.
270 Madison Avenue, New York, New York 10016

LIBRARY OF CONGRESS CATALOG CARD NUMBER: 73-82701
ISBN: 0-8247-6111-1
Current printing (last digit):
10 9 8 7 6 5 4 3 2 1

Printed in the United States of America

CONTENTS

Contributors v

Preface vii

1. APPLICATIONS OF THERMOMETRIC TITRIMETRY TO ANALYTICAL CHEMISTRY 1

    Lee D. Hansen, Reed M. Izatt, and James J. Christensen

    I. Introduction 2
    II. Theory 7
    III. Analytical Applications 28
    IV. Determination of the Stoichiometry of a Reaction 45
    V. Calorimetric Equilibrium Constant Determination 48
    VI. Equipment 62
    References 79

2. SOME UNUSUAL END-POINT DETECTION METHODS INVOLVING HETEROGENEOUS PROCESSES 91

    D. J. Curran

    I. Introduction 92
    II. Pressuremetric Titrations 93
    III. Cryoscopic Titrations 123
    IV. Phase Titrations 135
    V. Flame Photometric Titrations 172
    References 178

Author Index 183

Subject Index 191

CONTRIBUTORS

JAMES J. CHRISTENSEN, *Department of Chemical Engineering, Brigham Young University, Provo, Utah*

D. J. CURRAN, *Department of Chemistry, University of Massachusetts, Amherst, Massachusetts*

LEE D. HANSEN, *Center for Thermochemical Studies, Brigham Young University, Provo, Utah*

REED M. IZATT, *Department of Chemistry, Brigham Young University, Provo, Utah*

PREFACE

End-point determination is the very essence of titrimetric analysis: identification of the experimental situation when the amount of titrant corresponds to a theoretical equivalence point is the basic methodological principle. Thus, the birth of tit-trimetry probably coincides with the discovery that litmus could function as an acid-base indicator. Advances in the first half of this century included ingenious new chemical color indicators (e.g., redox indicators and metallochromic indicators) and instrumental methods of end-point detection. Indeed, potentiometric titrations have revolutionized volumetric analysis in the thirties.

This volume is devoted to developments which are likely to have a significant impact on the titrimetry of the seventies. Thermometric titrations are authoritatively reviewed by a group of authors who have made numerous pioneering contributions to the field. Several interesting and unconventional end-point detection methods are discussed by the inventor of pressuremetric titrations: they include cryoscopic titrations, phase titrations and flame photometric titrations, whose common denominator is the involvement of heterogeneous processes.

The two chapters of this book have been written by leading experts for an audience of practicing analytical chemists. The aim was to provide a concise, lucid, and topically coherent presentation of important innovations which have hitherto been scattered in journal articles and specialized monographs.

Joseph Jordan

New Developments in

TITRIMETRY

CHAPTER 1

APPLICATIONS OF THERMOMETRIC
TITRIMETRY TO ANALYTICAL CHEMISTRY*

Lee D. Hansen, Reed M. Izatt,
and James J. Christensen

Departments of Chemistry
and Chemical Engineering
Center for Thermochemical Studies
Brigham Young University
Provo, Utah

I. INTRODUCTION 2
   A. General 2
   B. Historical Review 6
II. THEORY 7
   A. Thermograms 7
   B. End-Point Determination 19
   C. End-Point Sharpness 23
   D. Direct Use of Reaction Enthalpies
      for Analysis 25
III. ANALYTICAL APPLICATIONS 28
   A. Aqueous Acid Base, Precipitation, and
      Metal Complexation Reactions 28

*Contribution No. 1 from the Center for Thermochemical Studies

	B. Aqueous Redox Reactions 34
	C. Reactions in Organic Solvents 39
	D. Reactions in Molten Salt Solvents 44
IV.	DETERMINATION OF THE STOICHIOMETRY OF A REACTION 45
V.	CALORIMETRIC EQUILIBRIUM CONSTANT DETERMINATION 48
	A. Theory 48
	B. Chemical Systems Studied by the Calorimetric Titration Method 53
VI.	EQUIPMENT 62
	A. Design and Construction of Thermometric Titration Equipment 62
	B. Description of Commercially Available Titration Equipment 63
	C. Construction of a Simple Inexpensive Thermometric Titrator 68
	REFERENCES 79

I. INTRODUCTION

A General

Extensive use has been made of the thermometric titration procedure both as an analytical tool for end-point or equivalence-point determination and as a calorimetric device for determining heats of reaction, heats of solution, heats of dilution, etc. Thermometric titrimetry is the technique in which the temperature of a reacting system is measured as a function of added titrant. The titrate (material titrated) and titrant may be liquids, gases, or suspended solids; the temperature change is produced by chemical reaction between

1. THERMOMETRIC TITRIMETRY

the titrate and titrant. The resultant temperature versus volume of titrant data can be evaluated analytically with respect to quantity or concentration of titrate and calorimetrically with respect to heat of reaction, heat of solution, etc. This chapter will be concerned only with the analytical aspects of thermometric titrimetry except where the heat of reaction has a bearing on the analytical application. A discussion of the calorimetric aspects of the method has been published [1].

Thermometric titrimetry, as opposed to most available analytical methods, depends on the enthalpy change (ΔH) for a reaction and not on the value of the Gibbs free energy change (ΔG) so long as ΔG is large enough to ensure a usable end point for the reaction. Thus, the method is applicable to many chemical systems where analytical methods such as EMF, spectrometry, and pH measurement fail or yield poor results. For many reactions which do not have large ΔG values, thermometric titrimetry provides a convenient and rapid method for calculating from a single determination changes in the free energy and entropy in addition to the change in enthalpy. Also, by noting end points and/or slope changes on the curve a check is obtained on the reactions assumed to be occurring. (The thermometric titration procedure produces temperature versus volume of titrant data which may be recorded in various ways, i.e., on a strip chart recorder, digital printout at timed intervals, etc. In any case, the resulting data may be plotted to produce a curve.)

Thermometric titrimetry permits the accurate and rapid quantitative determination of many substances whose concentrations were previously impossible or difficult to estimate. An example is seen in Fig. 1 where potentiometric and thermometric titration curves are compared for the titration of a strong

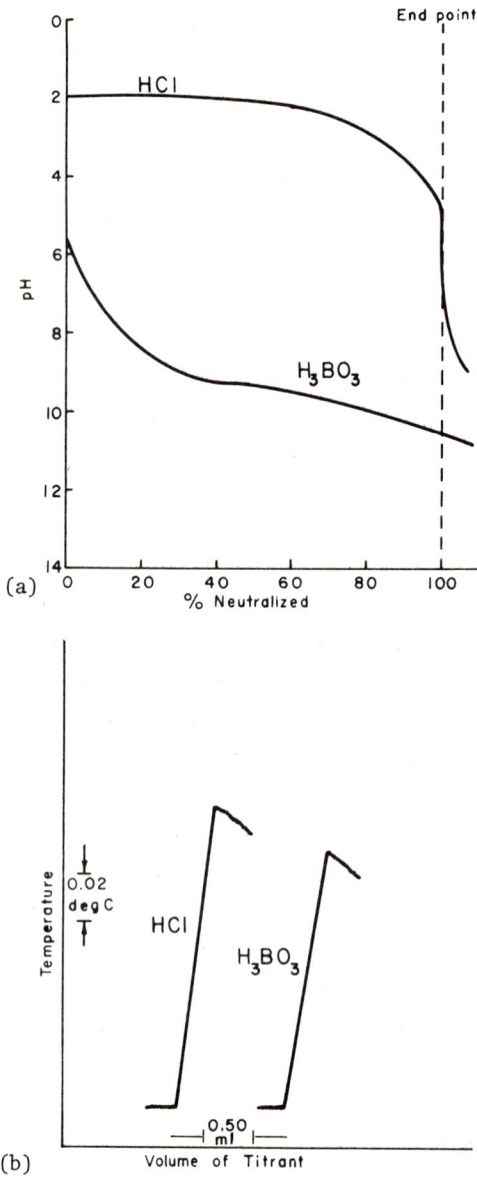

FIG. 1. POTENTIOMETRIC (a) AND THERMOMETRIC (b) TITRATIONS OF 0.01 M AQUEOUS SOLUTIONS OF HCl AND H_3BO_3 WITH NaOH. REPRINTED FROM [18, p. A12] BY COURTESY OF THE DIVISION OF CHEMICAL EDUCATION OF THE AMERICAN CHEMICAL SOCIETY.

1. THERMOMETRIC TITRIMETRY

acid (HCl) and a weak acid (H_3BO_3, pK = 9.24) [2] with a strong base (NaOH). The potentiometric titration curves show a well-defined inflection point in the case of HCl, but none in the case of boric acid. In contrast, the thermometric titration curves show well-defined end points for both HCl and H_3BO_3 since the temperature change is essentially independent of the pH of the solution.

Numerous investigators have shown that sharp end points are obtainable for a variety of systems and solvents. (A listing of chemical systems that have been studied by thermometric titrimetry is given in Ref. [3].) The method has also been shown to be applicable to end-point determinations for reactions occurring in emulsions and slurries [4-6]. Because thermometric titrimetry is independent of many properties of the solutions used (e.g., viscosity, optical clarity, dielectric constant), it is applicable to studies of gaseous [7,8] and nonaqueous [3, p. 139] as well as aqueous systems.

The thermometric titration procedures described in the literature can be classified as continuous or incremental depending on the mode of titrant addition. The continuous method produces continuous temperature versus volume of titrant data which may be recorded on a strip chart recorder. The incremental method gives a series of points, each one corresponding to the addition of a portion of the titrant. In much of the following discussion we will refer to the curves (thermograms) resulting from the continuous addition of titrant; however, it should be remembered that similar curves can be constructed from incremental titration data. The curves obtained from the continuous method are more convenient for determining end points and checking stoichiometry than those obtained from the incremental method; however, for slow reactions (i.e., reactions with reaction times longer than the equilibration time of the reaction vessel) the continuous method gives erroneous results and the incremental method must be used.

B. Historical Review

In 1913, Bell and Cowell [9] performed a thermometric titration by the laborious process of adding titrant incrementally from a buret and measuring the temperature rise after each addition with a Beckman thermometer. Thermograms obtained in this manner can be analyzed either for the concentration of one of the reactants if the stoichiometry is known or for the stoichiometry if the concentrations of all the reactants are known. However, the crudeness of the method and equipment gives rise to many difficulties, the more serious of which are (a) the length of the experiment due to the incremental addition and (b) erroneous slopes or inflection points resulting from the large response lags of the thermometer and heat transfer between the solution and the surroundings. It was not until the introduction of a rapid response temperature sensing device (thermistor) [4,10], together with an electronic recorder and automatic continuous delivery buret [11] that the method became a rapid, effective analytical tool. Further developments have led to the use of a differential thermistor circuit with twin titration vessels in an attempt to reduce the errors due to heat losses and heat of dilution corrections [12,13] and to circuits for differentiating the thermograms to give first and second derivatives for a more accurate estimation of equivalence points [14-16].

Several review articles, books, and chapters are available describing the use of thermometric titrimetry as an analytical tool including descriptions of equipment and procedures [1,3,7,17-34].

1. THERMOMETRIC TITRIMETRY

In the following sections of this presentation, the theory of thermometric titrimetry is discussed, examples illustrating the analytical applications of the method are given, and the design, construction, and availability of thermometric titration equipment are described.

II. THEORY

A. Thermograms

A complete thermogram (Fig. 2) consists of four regions: (a) a prereaction region where no titrant is added and where the titrate temperature is nearly invariant with time; (b) a reaction period in which the change of the temperature of the solution with time is largely a result of the reaction(s) occurring in the calorimeter; (c) a postreaction region in which the temperature changes slightly with time due to continued addition of titrant; and (d) a postreaction region where no titrant is added and the temperature again is nearly invariant with time. The intersection of the prereaction and reaction period lines (x) is the starting point of the titration while the intersection of the reaction and postreaction period lines (y) marks the end of the chemical reaction.

During the reaction period (Fig. 2, region b), the thermogram usually deviates from a straight line with its form being determined by (a) the number of reactions occurring, (b) the relative magnitudes of the equilibrium constants for the reactions, (c) the relative magnitudes of the heats of reaction, (d) the relative titrant and titrate concentrations, (e) the titrant delivery rate, (f) the change of heat capacity during the titration, and (g) the heat effects due to any

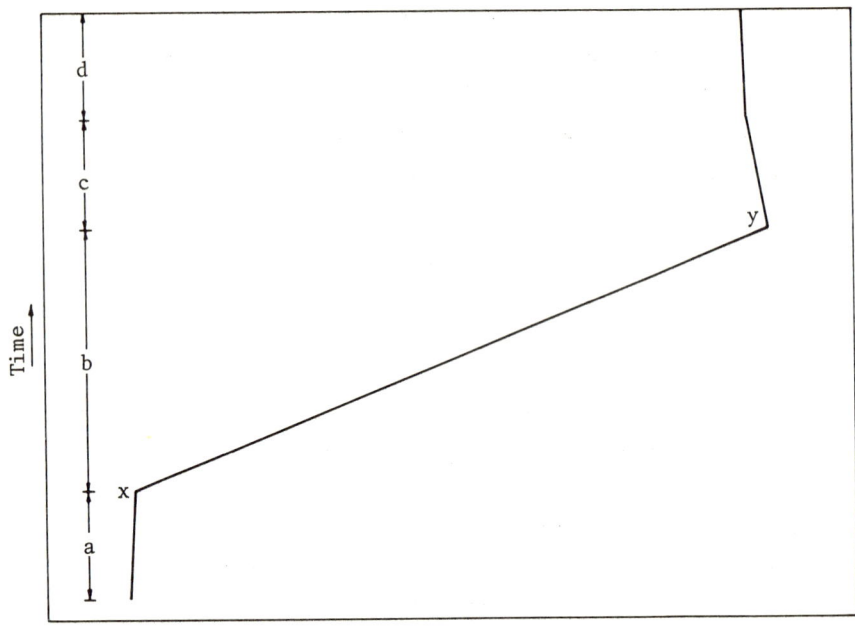

FIG. 2. COMPLETE THERMOGRAM: (a) PREREACTION REGION, (b) REACTION REGION, (c) POSTREACTION REGION WITH CONTINUED TITRANT DELIVERY, (d) POSTREACTION REGION.

temperature difference between the titrant and the titrate solutions, heat losses, heats of dilution, etc. (This is based on the assumption that all reactions occurring are rapid.)

Factor (a) determines the maximum number of end points that can be obtained on the thermogram; factors (b) and (c), however, determine the actual number of end points observed. The relative magnitudes of the consecutive equilibrium constants determine whether the end points are sharp or rounded. For consecutive reactions, sharp end points will be observed when the difference between the logarithms of two consecutive equilibrium constants is greater than ∿3. The relative magnitudes of the heats of reaction determine for multiple end points

1. THERMOMETRIC TITRIMETRY

whether the slope of the thermogram will increase or decrease at the end points (i.e., if $-\Delta H_2 > -\Delta H_1$ the slope will increase, that is, the temperature rise will become greater for the same amount of titrant added and if $-\Delta H_2 < -\Delta H_1$, the slope will decrease, assuming ΔH_1 and ΔH_2 are the only heats determining the shape of the thermogram). In the special case where the consecutive heats of reaction are equal within the limits of measurement of the equipment, no break point will be observed regardless of the relative magnitudes of the equilibrium constants.

Factors (d) and (e) influence the form of the thermogram with respect to the temperature rise obtained in a given time interval. Since the titrant is delivered at a constant rate, the thermogram can be considered to be a plot of either the amount of titrant added or the time versus temperature. Low titrant concentration and slow titrant delivery rates give less reaction and, therefore, less temperature rise per unit time than do high concentrations and fast delivery rates. Thus, it is possible to extend or compress the time scale with respect to the temperature scale and the amount of reaction taking place, which is useful when it is desirable to investigate one portion of the thermogram.

Factor (f) is present in the usual thermometric titration since mass must of necessity cross the boundary of the thermodynamic system, thus increasing the heat capacity of the system. A given increment of heat produced near the end of the titration does not cause as large a temperature change as does the same increment near the beginning of the titration. This effect can be minimized by keeping the ratio of titrant volume to titrate volume small.

Factors (d)-(g) are always present in an adiabatic thermometric titration but are relatively unimportant in any discussion of the chemistry of the system under investigation.

Let us now examine seven representative thermograms and species distribution curves for selected acid-base and metal ligand systems, at the same time remembering that the shapes of the thermograms are primarily determined by factors (a)-(c).

1. *Sodium Malonate-HClO₄ System*

Figure 3 shows a thermogram and a species distribution for the titration of a sodium malonate (Na₂A) solution with HClO₄ [35].. The sharp end point upon addition of one equivalent of HClO₄ corresponds to the nearly stoichiometric formation of HA⁻. This end point is a result of (a) the large difference between the pK values (pK_{HA} = 5.7 and pK_{H_2A} = 2.8 [2]) and (b) the difference in the heats of protonation of A^{2-}

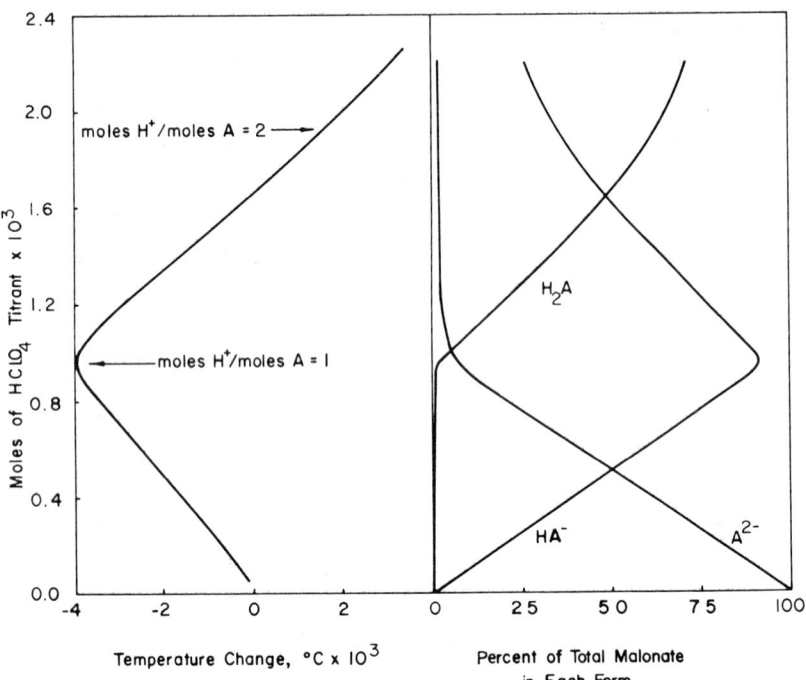

FIG. 3. THERMOGRAM AND SPECIES DISTRIBUTION FOR THE TITRATION OF SODIUM MALONATE, Na₂A, WITH HClO₄.

1. THERMOMETRIC TITRIMETRY

and HA (0.92 and -0.29 kcal/mole, respectively). The almost complete absence of a second end point is a result of the relatively small pK value for H_2A dissociation.

2. *Sodium Suberate-HClO₄ System*

Figure 4 shows the thermogram and species distribution for the titration of a sodium suberate (Na_2A) solution with $HClO_4$ [35]. Even though the ΔH values for consecutive addition of H^+ to A^{2-} (0.64 and 0.39 kcal/mole, respectively) differ, a sharp end point between the formation of HA^- and H_2A is not observed since the pK values are within one pK unit of each other (pK_{HA} = 5.4 and pK_{H_2A} = 4.5). However, the pK_{H_2A} value for suberic acid, in contrast with that for malonic acid is large enough that a well-defined end point is seen upon

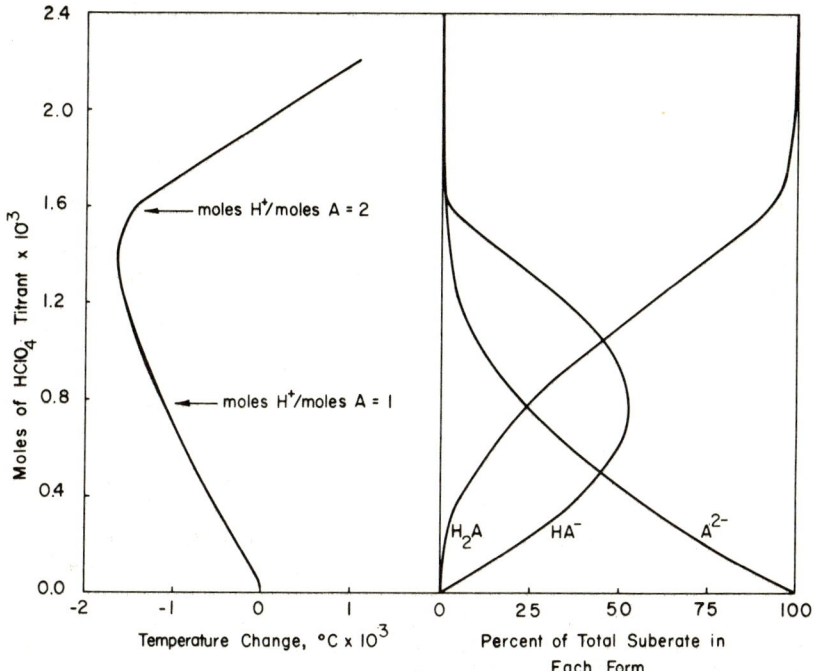

FIG. 4. THERMOGRAM AND SPECIES DISTRIBUTION FOR THE TITRATION OF SODIUM SUBERATE, Na_2A, WITH $HClO_4$.

stoichiometric formation of H₂A. This end point could be used to determine the total number of equivalents of A²⁻ and HA⁻ initially present.

3. *Adenosinediphosphate-NaOH System*

In Fig. 5 is shown the thermogram and species distribution for the titration of an adenosinediphosphate ion (ADP) solution with NaOH [36]. The zwitterion form of ADP (I) used in this titration has three protons which are titrated by the NaOH in 1,2,3 order as shown. The shape of the thermogram reflects two effects. First, the temperature rise at any point along the curve is a result of the simultaneous reaction of protons from two ADP species with the NaOH titrant although

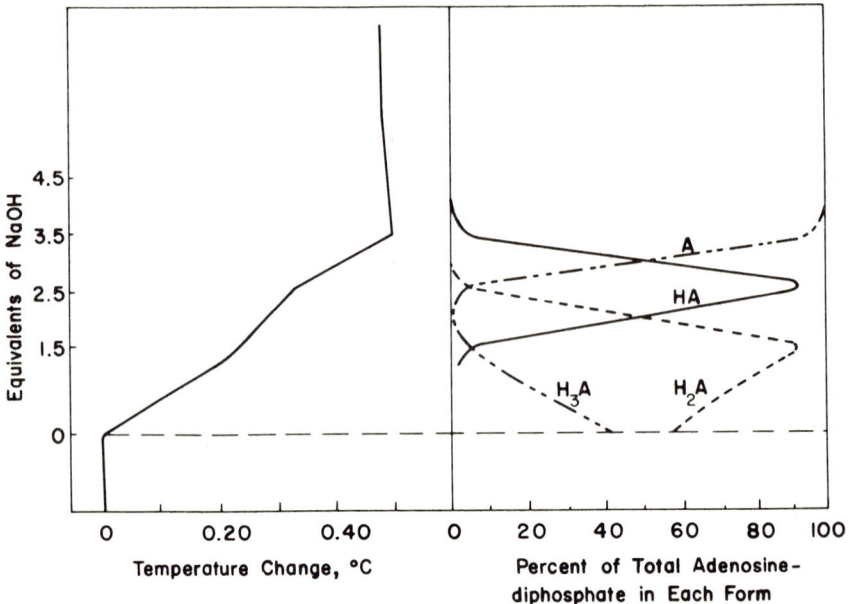

FIG. 5. THERMOGRAM AND SPECIES DISTRIBUTION FOR TITRATION OF ADENOSINEDIPHOSPHATE ION, H₃A, WITH NaOH. REPRINTED FROM [36, p. 1031] BY COURTESY OF THE AMERICAN CHEMICAL SOCIETY.

1. THERMOMETRIC TITRIMETRY

(I)

the contribution is mainly from one species in each case. The result is a nonlinear thermogram with rounded end points. Second, the ΔH value for proton ionization from site 2 (4.1 kcal/mole) is quite different from that from sites 1 and 3 (∼-1 kcal/mole) which explains the S-shaped thermogram (pK_1 = 1.2, pK_2 = 4.2, and pK_3 = 7.0 [2]). The sharp end point corresponds to the stoichiometric removal of protons from site 3, i.e., from HA^{2-}.

4. *Sodium Glycinate-HClO4 System*

Figure 6 gives the thermogram and species distribution for the titration of glycinate ion with $HClO_4$ [37]. The reactions which take place are, successively, $A^- + H^+ = HA^{\pm}$ (log K = 9.8, $\Delta H°$ = -10.57 kcal/mole [2]) and $HA^{\pm} + H^+ = H_2A^+$ (log K = 2.4, ΔH = -0.98 kcal/mole [2]). The thermogram is interesting in that it has both a sharp and a rounded end point illustrating the effect of the relative magnitudes of the pK values on the end points.

5. *Glycine Hydrochloride-NaOH System*

In Fig. 7 is shown the thermogram and species distribution for the titration of glycine hydrochloride with NaOH [37, 38]. The successive reactions occurring in this titration are

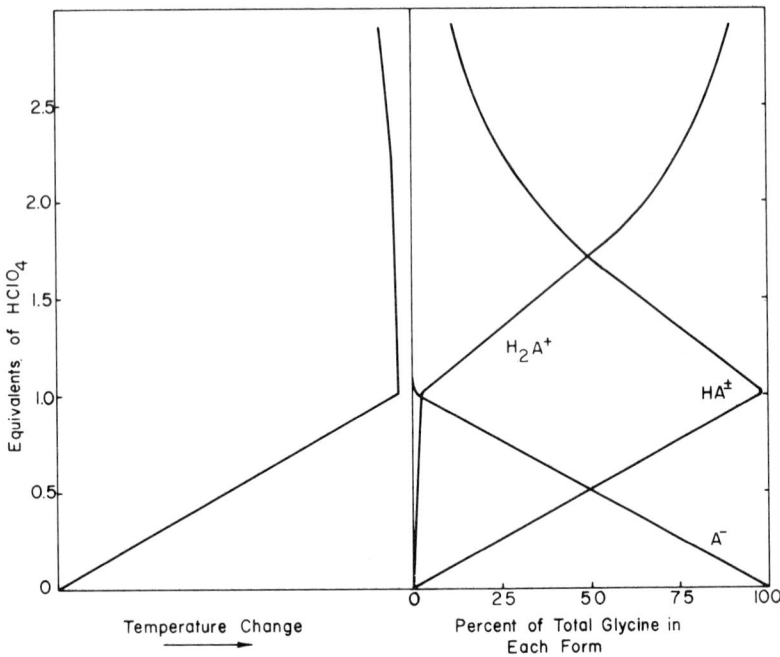

FIG. 6. THERMOGRAM AND SPECIES DISTRIBUTION FOR TITRATION OF SODIUM GLYCINATE, NaA, WITH HClO$_4$.

$H_2A^+ + OH^- = HA^{\pm} + H_2O$ (log K = 11.6, ΔH = -12.4 kcal/mole [2]), and $HA^{\pm} + OH^- = A^- + H_2O$ (log K = 4.2, ΔH = -2.8 kcal/mole [2]). The thermogram is seen to be similar to the one given in Fig. 6. This similarity arises because in each case the log K value for the first reaction is large compared to that for the second reaction making the first reaction quantitative. In addition, the ΔH values for the first reaction are similar, as are those for the second reaction. The sharper second end point in Fig. 7 compared to that in Fig. 6 results from the larger log K value in the case of the second reaction in Fig. 7.

1. THERMOMETRIC TITRIMETRY

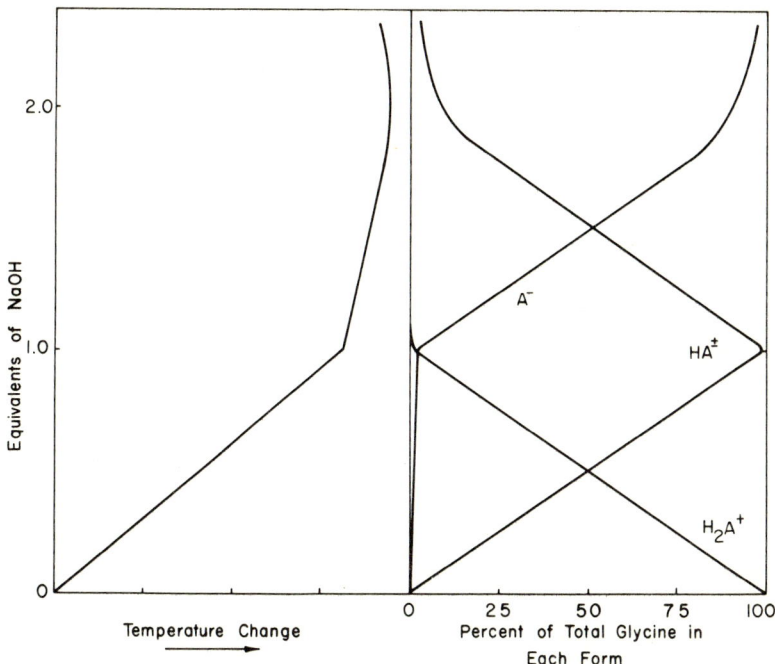

FIG. 7. THERMOGRAM AND SPECIES DISTRIBUTION FOR TITRATION OF GLYCINE HYDROCHLORIDE, H_2A^+, WITH NaOH.

The thermograms in Figs. 6 and 7 illustrate how it was possible to obtain clear end points for both the A^- and H_2A species by changing the titrant. This shows that it may be possible, by a judicious choice of titrant, to obtain clear end points for any one of the reactions occurring in a multiple reaction system.

6. *Mercury(II) Perchlorate-NaCl System*

Figure 8 shows the thermogram and species distribution for the titration of a $Hg(ClO_4)_2$ solution with NaCl [39]. A

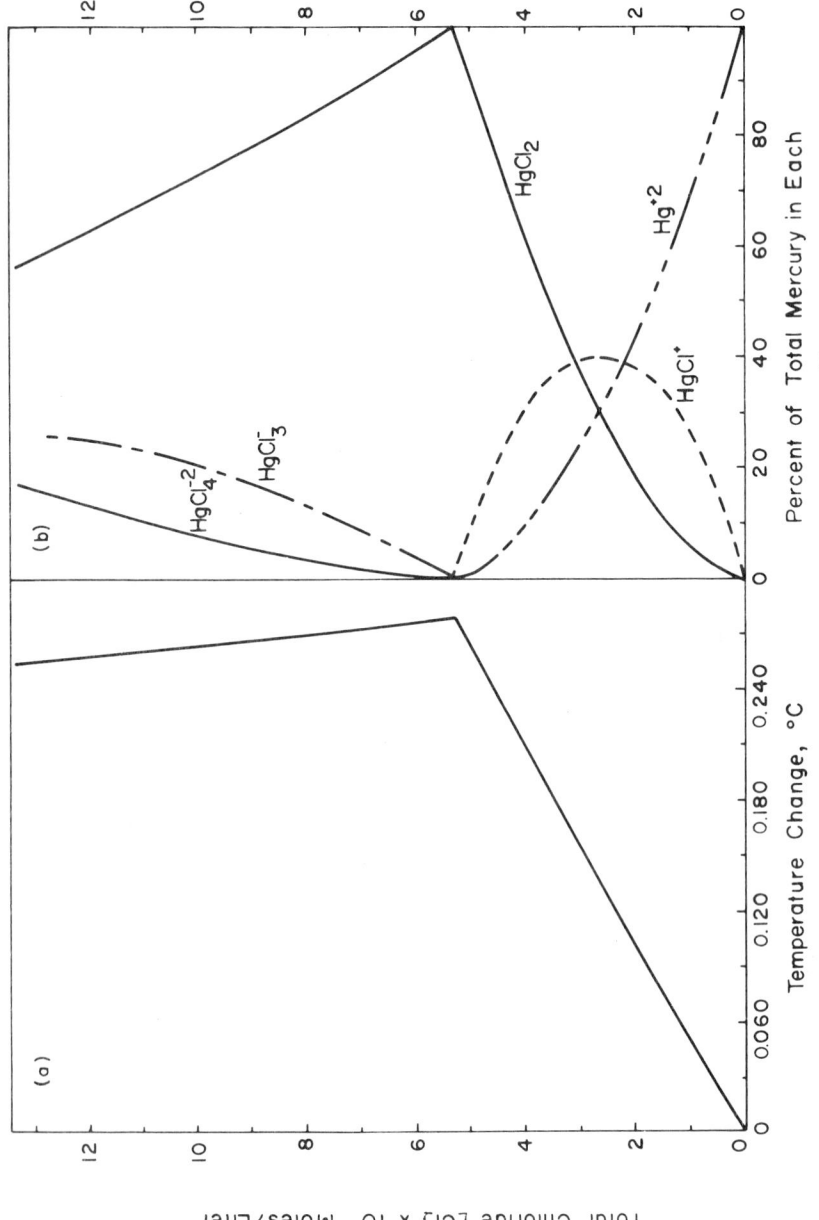

FIG. 8 THERMOGRAM AND SPECIES DISTRIBUTION FOR TITRATION OF $Hg(ClO_4)_2$ WITH NaCl. REPRINTED FROM [39, p. 131] BY COURTESY OF THE AMERICAN CHEMICAL SOCIETY.

1. THERMOMETRIC TITRIMETRY

single end point corresponding to the formation of $HgCl_2(aq)$ is observed in this system. The absence of an end point corresponding to the formation of $HgCl^+$ may be attributed to the very small difference between the consecutive constants for the formation of $HgCl_2(aq)$ (log K_{HgCl^+} = 6.62 and log K_{HgCl_2} = 6.36 [40]). Since the consecutive ΔH values are not equal (-5.5 and -7.3 kcal/mole, respectively) there is a slight curvature of the thermogram prior to the end point. No end point is observed corresponding to $HgCl_3^-$ or $HgCl_4^{2-}$ formation because of the small constants for their formation from $HgCl_2$ (aq) (log $K_{HgCl_3^-}$ = 0.95 and log $K_{HgCl_4^{2-}}$ = 1.05 [41]).

7. *Mercury(II) Perchlorate-NaCN System*

A thermogram and species distribution are shown in Fig. 9 [1, 42] for the titration with NaCN of a solution of mercuric perchlorate containing sufficient perchloric acid to repress hydrolysis. Because of the relative magnitudes of the equilibrium constants of the cyanide-containing species, the thermometric titration curve shows three distinct reaction regions. In region 1 as NaCN titrant is added to the acidic Hg^{2+} solution, $HgCN^+$ and $Hg(CN)_2(aq)$ are formed since the constants for their consecutive formation (log K_{HgCN^+} = 17.00, log $K_{Hg(CN)_2}$ = 15.75) are much larger than those of any of the other possible species. A sharp end point is observed upon addition of two equivalents of NaCN indicating stoichiometric formation of $Hg(CN)_2(aq)$. In region 2 continued addition of NaCN results in a marked slope change of the curve followed by a sharp end point which is identified with the stoichiometric disappearance of the original excess H^+ indicating formation of HCN (pK = 9.21). Region 3 corresponds to the consecutive formation of $Hg(CN)_3^-$ (log K = 3.56) and $Hg(CN)_4^{2-}$ (log K = 2.66) from $Hg(CN)_2(aq)$.

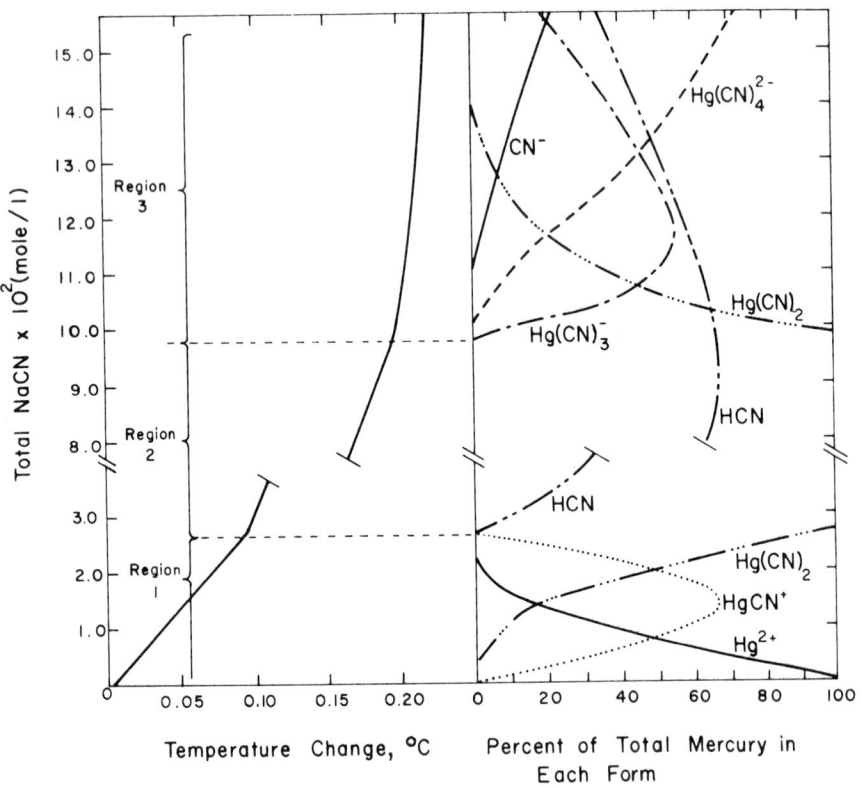

FIG. 9. THERMOGRAM AND SPECIES DISTRIBUTION FOR TITRATION OF $Hg(ClO_4)_2$, $HClO_4$ WITH NaCN. REPRINTED FROM [1, p. 562] BY COURTESY OF WILEY (INTERSCIENCE) PUBLISHERS.

In summary, the above examples indicate how variations in the number of reactions and in the relative magnitudes of the corresponding equilibrium constants and heats determine the existence or absence of analytically useful end points.

1. THERMOMETRIC TITRIMETRY

B. End-Point Determination

The usual method of determining the end point of a thermometric titration is to extrapolate the reaction and post-reaction period lines to their intersection point (point y, Fig. 2). This extrapolation is usually short and causes little error. However, if a reaction is not quantitative, the thermogram will be considerably rounded in the equivalence point region and any extrapolation will be uncertain. In these cases it is sometimes helpful to use first or second derivative curves to find the end points. These curves, illustrated in Fig. 10, are obtained by electronic differentiation of the output signal from the temperature sensing circuit. Circuits for this purpose have been described by several workers [14-16]

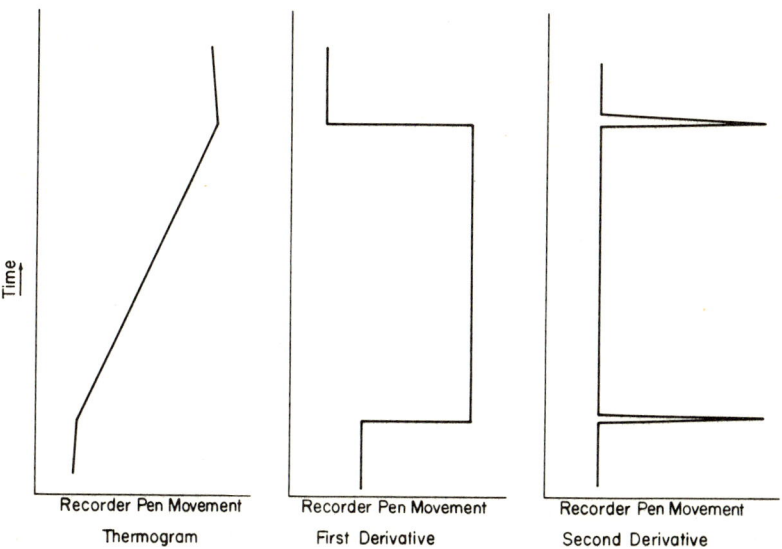

FIG. 10. FIRST AND SECOND DERIVATIVE CURVES OF A THERMOGRAM.

and, as has been shown by Priestly [15], the derivative curves can be used for automatic end-point detection. This development makes thermometric titrimetry a suitable method for process control and remote control applications.

When end-point detection is difficult because of a small ΔH value for the reaction being studied, two approaches may be used: (a) amplification of the signal from the temperature sensor and (b) addition of a "thermochemical indicator." Approach (a) is particularly useful when it is desired to locate an end point between two consecutive reactions, a situation in which the second approach is usually not feasible. The use of amplification to sharpen the end point is illustrated in Fig. 11 for the titration of sodium malonate (Na_2A) with $HClO_4$ [35]. Curve (a) is the result of amplifying curve (b) by a factor of 20. Amplifying the temperature signal greatly expands that axis of the graph, increases both slopes, and accentuates the differences between the slopes making the end point sharper.

In approach (b) a second reaction (involving the thermochemical indicator) is initiated at the end point of the reaction being studied. For example in the case of the titration of acetate ion with a strong acid, ΔH = -0.01 ± 0.01 kcal/mole [43], no end point is observed. With the addition to the mixture of a thermochemical indicator such as SO_4^{2-}, which is a weaker base ($pK_{HSO_4^-}$ = 1.97 [44]) than acetate ion ($pK_{acetic\ acid}$ = 4.766 [35]), but which has a large heat of reaction with hydrogen ion (∼5 kcal/mole [44]), an end point will be observed as seen in Fig. 12 [45]. Sodium sulfate is a good thermochemical indicator for the titration of basic substances such as carboxylates and phosphates with strong acids where ΔH° values for the interaction are near zero [35,36]. The reactions of many amine salts with strong base also have ΔH values near zero [46]. Since the pK values of the protonated

1. THERMOMETRIC TITRIMETRY   21

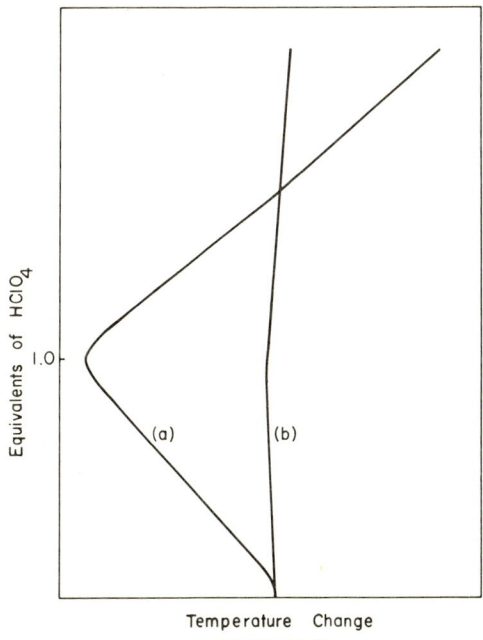

FIG. 11. EFFECT OF AMPLIFYING THE THERMOGRAM FOR TITRATION OF SODIUM MALONATE, Na$_2$A, WITH HClO$_4$. TEMPERATURE READING AMPLIFIED BY FACTOR OF 20 (a) AND 1 (b).

amine salts are generally 9-10.5 [46], a possible thermochemical indicator in these cases would be glucose (pK = 12.46, $\Delta H°$ = 7.7 kcal/mole for proton ionization [47]).

Another variation on the use of thermochemical indicators is to add to the titrate solution a pair of reactants which require the presence of a small amount of titrant species to catalyze their reaction. The reaction of Ce(IV) with As(III), for example, is catalyzed by traces of iodide ion and this redox reaction can be used to indicate the end point for titrations with iodide ion [48-50]. In a similar fashion, the decomposition of H$_2$O$_2$ or the reaction of H$_2$O$_2$ with resorcinol, both of which are catalyzed by Mn^{2+}, can be used to indicate

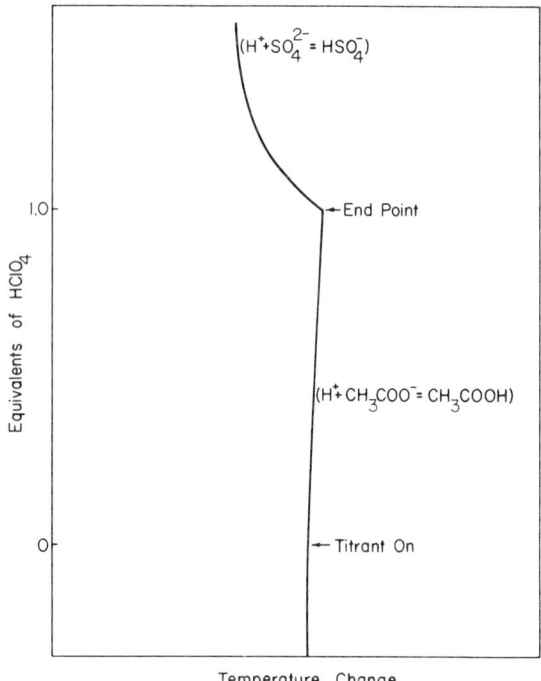

FIG. 12. THERMOGRAM FOR TITRATION OF A MIXTURE OF SODIUM ACETATE AND SODIUM SULFATE WITH HClO$_4$.

the end point in titrations involving Mn^{2+} [49,51]. The reaction of a nonaqueous solvent with itself as catalyzed by excess titrant provides still another variation. Acetone undergoes an aldol condensation in the presence of OH$^-$ and this reaction has been used to indicate the end point for acid-base reactions in acetone [6,52].

Difficulties in end-point determination are also often encountered with nonaqueous solvent systems or very concentrated solutions where the ΔH value for the reaction of interest is much smaller than the heat of dilution of the titrant [12]. In this event, the total temperature rise may be so

1. THERMOMETRIC TITRIMETRY

large that the sensitivity of the temperature sensor must be reduced to the point that the heat effect of interest is too small to be observed. This problem can be reduced by performing a differential titration using twin calorimeters and two burets. In this kind of titration the two burets contain identical titrant solutions, but only one calorimeter vessel contains the reactant of interest. The differential temperature sensor is usually a Wheatstone bridge with thermistors in two opposite arms, one in each calorimeter vessel. Thus, the temperature effect due to dilution of the titrant is essentially cancelled and only the effect due to the reaction of interest is recorded [12]. An example of the use of this method is the titration in an anhydrous acetic acid solvent of very weak bases with $HClO_4$ [13].

The precision of end-point determination assigned by most investigators has been of the order of 0.5 to 1%. However, this precision may be improved to about 0.1% by greatly expanding the time and temperature scales near the end point [53]. This is conveniently and precisely done by first adding 90% of the titrant required and then making small incremental additions of titrant at maximum equipment sensitivity until the end point is passed. The difficulties of exactly locating the starting point of the thermogram, of lag time in the reaction or recording system, and of any variation in buret and recorder speeds are thus eliminated or greatly reduced.

C. End-Point Sharpness

A rounding in the end-point region can usually be attributed to an incomplete reaction resulting from a small equilibrium constant (log K < 4) for the reaction in the calorimeter. However, rounding may also be caused by a slow reaction, in which case a meaningful end point cannot be obtained by a simple extrapolation, but may sometimes be found by an iterative mathematical treatment of the equations describing the kinetics of the reaction [34,54].

If the rounded end point is due to a small equilibrium constant for the reaction, certain limitations are present, but meaningful analytical results can be obtained. A discussion of the limitations of the thermometric titration method as compared to other analytical methods commonly used for endpoint detection is given below.

Application of the usual equilibrium calculations show that at the equivalence point for the reaction R + T = P the fraction of titrant, T, reacted, is related by Eq. (1) to the equilibrium constant for the reaction, K_f, and the total concentration of R, C_R [55].

$$K_f C_R = \frac{\alpha}{(1 - \alpha)^2} \qquad (1)$$

Since α is also equal to the ratio of the temperature observed at the end point to the temperature which would have been observed if the reaction had gone to completion, it is closely related to the curvature at the end point. Values of α less than ~0.96 (corresponding to $K_f C_R$ < 600) will not, in general, give analytically useful end points because of the necessity of extrapolating the thermogram through more than one-half of the reaction region. This is, however, a very much lower value of $K_f C_R$ than can profitably be used in any of the logarithmic types of analytical determination (e.g., potentiometric titration in which the lower limit on α is ~0.999 corresponding to a $K_f C_R$ value of 10^6). The thermometric titration method therefore is of great advantage in applications involving weakly interacting systems or very dilute solutions where the sigmoid shape curves typically obtained from the logarithmic analytical methods become too flat for useful

1. THERMOMETRIC TITRIMETRY 25

end-point detection. The successful thermometric titration of boric acid solution with sodium hydroxide solution is an illustration of this point (see Fig. 1).

D. Direct Use of Reaction Enthalpies for Analysis

If the ΔH value of a reaction is known, it is possible to determine the amount of one of the reactants by adding an excess of the other reactant and measuring the heat evolved by the reaction. This principle is incorporated in two analytical methods which are related to thermometric titration, i.e. direct injection enthalpimetry (DIE), and continuous flow enthalpimetry (CFE).

DIE uses the above principle directly as stated and has been tested on several chemical reactions [56], i.e., HCl/NaOH, H_3BO_3/NaOH, Pb^{2+}/EDTA, Mg^{2+}/EDTA, where the unknown is listed first in each case. A precision of ±2% (relative standard deviation) is reported for all systems studied. Water in organic solvents can also be determined by DIE, the absorption of water on molecular sieves being used to produce the temperature pulse [57].

In CFE, two streams, each containing a reactant, are mixed and the temperature rise in the product stream due to reaction is measured [58]. Priestley [59] has described an ingenious thermistor circuit which measures the temperature difference between the product stream and the average of the temperatures of the reactant streams. CFE has obvious value in process control applications. A comparison of actual (nominal) concentrations with those determined by CFE is made in Table 1.

Another possible analytical use of the above principle is the determination, without separation, of substances which have identical (or very similar) equilibrium constants, but

TABLE 1[a]

Comparison of Observed and Nominal Values Using CFE Method of Analysis

Sample	Results					
Reagent, 0.6 M sodium hydroxide						
Nominal values	0.050	0.100	0.200	0.300	0.400	0.500
Hydrochloric acid	0.051	0.103	0.200	0.303	0.402	0.500
Nitric acid	0.050	0.100	0.200	0.300	0.396	0.500
Boric acid	0.049	0.099	0.196	0.302	0.400	0.500
Acetic acid	0.050	0.098	0.200	0.305	0.399	0.500
Reagent, 0.6 M hydrochloric acid						
Nominal values	0.025	0.050	0.100	0.150	0.200	0.300	0.400	0.500
Sodium hydroxide	--	0.050	0.101	--	0.200	0.299	0.400	0.500
Potassium hydroxide	--	0.049	0.103	--	0.200	0.303	0.399	0.500
Ammonia solution	--	0.051	0.100	--	0.199	0.300	0.396	0.500
Pyridine	--	0.055	0.099	--	0.201	0.299	0.401	0.500
Trisodium orthophosphate	0.025	0.050	0.102	0.150	0.200	--	--	--
Sodium sulfite	0.025	0.050	0.106	0.153	0.200	--	--	--
Reagent, M EDTA (Tetrasodium salt)						
Nominal values	0.100	0.200	0.300	0.400	0.500	0.600	0.700	0.800
Barium chloride	0.099	0.199	0.300	0.395	0.500	--	--	--
Copper sulfate	--	0.200	--	0.403	--	0.600	--	--
Lead acetate	--	0.198	--	0.397	--	0.597	0.700	--
Nickel chloride	--	0.202	--	0.395	--	0.598	--	0.800
Manganese sulfate	0.100	0.201	0.298	0.400	0.500	--	--	--

[a]Reprinted from Ref. [58, p. 592] by courtesy of the Society for Analytical Chemistry.

1. THERMOMETRIC TITRIMETRY

different ΔH values [45]. For two reactants having these properties, the relations between total heat produced by the reactions, Q, the total number of moles of titrant reacted, n_T, the number of moles of each reactant, n, and the molar enthalpy changes, ΔH, for the reactions are given in Eq. (2) and (3). Since n_T may be found from the one end point which will appear on the thermogram corresponding to the total reaction of both reactants and ΔH_1 and ΔH_2 may be determined in

$$Q = n_1 \Delta H_1 + n_2 \Delta H_2 \qquad (2)$$

$$n_T = n_1 + n_2 \qquad (3)$$

separate experiments on each of the pure reactants, the only unknowns are n_1 and n_2 which are easily calculated by the simultaneous solution of Eq. (2) and (3).

The method has been tested on mixtures of sodium acetate (pK_a (acetic acid) = 4.8, $\Delta H_i \sim 0$) and pyridine (pK_a (protonated pyridine) = 5.3, $\Delta H_i \sim 5$) and of phenol (pK_a = 10.0, $\Delta H_i \sim 6$) and glycine (pK_a = 9.8, $\Delta H_i \sim 11$) which were titrated with perchloric acid and sodium hydroxide, respectively [45]. The relative error in the determination of each component of the mixture was found to be about 5% for the millimolar amounts taken. An error analysis of Eq. (2) and (3) showed that the error to be expected in n_1 (or in n_2) was primarily dependent on the inverse of the difference between ΔH_1 and ΔH_2. This method should also be applicable to redox, precipitation, complexation, and Lewis acid-base reactions.

A modification of the above method was used by Wrathall et al. [60] to determine microconstants for proton ionization from the cysteine zwitterion ($HSCH_2CH(NH_3^+)COO^-$). In cysteine there are two proton dissociating groups, i.e., -SH and $-NH_3^+$. A pH titration curve of these groups with NaOH gives two buffer regions. Both -SH and $-NH_3^+$ groups ionize in each region, and Eq. (2) and (3) can be used to determine the ratio of $-NH_3^+$ to

-SH groups which ionize in each of the buffer regions. Using this method, Wrathall et al. [60] were able to calculate microconstants for the four proton ionization reactions of cysteine.

III. ANALYTICAL APPLICATIONS

The analytical applications of thermometric titrimetry have been classified in this discussion as (a) aqueous acid-base, precipitation, and metal complexation reactions, (b) aqueous redox reactions, (c) reactions in organic solvents, and (d) reactions in molten salt solvents. The accuracy claimed for the determinations in the systems discussed is usually better than 1%.

A. Aqueous Acid-Base, Precipitation, and Metal Complexation Reactions

The most common analytical use of thermometric titrimetry has been the determination of acids or bases in aqueous solutions.

Jordan and Dumbaugh [61] demonstrated that sodium hydroxide solution could be used to determine boric (pK = 9.24), acetic (pK = 4.74), monochloroacetic (pK = 2.80), and trichloroacetic (pK = 0.52) acids in aqueous solution. The pK range covered by these acids shows the wide range of applicability of the method. Thermometric titration has also been used to analyze and identify mixtures of amino acids by use of their acid-base reactions [38].

DeLeo and Stern [62] used HCl, NaOH, and $AgNO_3$ titrants to determine the pharmaceutically important compounds chloropheniramine maleate, theophylline, aminophylline, nicotinamide, and chloropromazine hydrochloride. The procedures developed by them for determining these compounds proved to be faster,

1. THERMOMETRIC TITRIMETRY

simpler, and at least as accurate as existing standard methods.

Thermometric titrimetry has been applied to the analysis of detergents and dilute detergent solutions which are difficult to analyze by most methods because of the formation of micelles. Two test compounds have been examined: 2-dodecylbenzenesulfonate as an example of an alkylbenzenesulfonate (ABS) type detergent and cetylpyridinium chloride as an example of an invert detergent. Both substances and the classes of compounds they represent are of significance in sewage disposal and pollution problems. Jordan et al. [63] titrated 2-dodecylbenzenesulfonate solutions to a sharp end point with a solution of Hyamine 1622, a quaternary amine. They reported that the method was fast and had an accuracy of better than 1% at ABS concentrations between 4 and 17 mM. The results obtained by Weiner and Felmeister [64] for titration of cetylpyridinium chloride solution with a solution of Orange II were similarly good.

Tyson et al. [12] studied several reactions using HCl solutions as titrants in a twin thermometric titration calorimeter. The data collected in that study are given in Table 2.

Priestley [15] tested the use of HCl, Na$_4$EDTA, and sodium carbonate solutions as titrants for a large number of different compounds as shown in Table 3. Nitrilotriacetate, diaminocyclohexanetetraacetate, citrate, tartrate, oxalate, and malonate ions have also been used as chelating titrants for a variety of metal ions [53,65-69].

The thermometric titration technique has been used to determine the composition of mixtures without prior separation where pK values and the heats of reaction were sufficiently different to give an end point for each reaction [see Eq. (1) and discussion thereof]. Miller and Thomason used NaOH titrant to determine H$_3$BO$_3$ in H$_2$SO$_4$- H$_3$BO$_3$ mixtures [70] and total acid in the presence of fluoride ion and the hydrolyzable

TABLE 2a

Results of Analytical
Determinations Using Thermometric Titration

Substance titrated	Titrant (N HCl)	Amount added (meq)[b]	Amount found (meq)	Mean deviation (%)	Error (%)
AgNO$_3$	1.013	0.996(4)	0.997	0.1	+0.1
		0.4989(4)	0.499$_6$	0.2	+0.1
		0.1989(3)	0.202$_8$	0.7	+1.9
AgNO$_3$	0.9698	1.005(5)	1.003	0.4	-0.2
		0.2987(10)	0.298$_8$	0.5	0.0
NaOH	1.013	2.296(3)	2.275	0.1	-0.9
		1.150(6)	1.140	0.1	-0.9
		0.4906(3)	0.484$_8$	0.0	-1.2
		0.4586(5)	0.454$_2$	0.2	-1.0
NaOH	0.9698	1.643(3)	1.624	0.0$_4$	-1.1
		0.3289(4)	0.325$_4$	0.2	-1.1
NH$_3$	0.9698	0.3932(5)	0.383$_3$	0.8	-2.5
Pyridine	0.9698	0.991(5)	0.983	0.2	-0.8
		0.3953(4)	0.393$_9$	0.2	-0.4
Ethylenediamine	0.9698				
Overall		0.5024(5)	0.505$_4$	0.1	+0.6
1st NH$_2$		0.2512(5)	0.253$_9$	1.9	+1.1
2nd NH$_2$		0.2512(5)	0.251$_5$	2.0	+0.1
Ethanolamine	0.9698	0.3987(4)	0.397$_5$	0.9	-0.3
2-Amino-1-butanol	0.9698	0.4027(9)	0.400$_2$	0.9	-0.6
2-Methyl-2-amino-1-propanol	0.9698	0.4009(8)	0.399$_2$	0.6	-0.4
2-Methyl-2-amino-1,3-propanediol	0.9698	0.3982(6)	0.399$_3$	0.5	+0.3
Tris(hydroxymethyl)-aminomethane	0.9698	0.3987(6)	0.399$_1$	0.4	+0.1
		0.3991(7)	0.398$_3$	1.2	-0.2

[a]Reprinted from [12, p. 1643] by courtesy of the American chemical Society.
[b]Numbers in parentheses represent the number of duplicate determinations.

1. THERMOMETRIC TITRIMETRY

TABLE 3[a]

Results of Analytical
Determinations Using Thermometric Titration

Titrate	C_m	No. of titrations	Sigma (%)[b]
\multicolumn{4}{c}{Titrations with M hydrochloric acid as titrant}
NaOH	$M/60$	10	0.6
AgNO₃	$M/60$	10	0.5
NH₄OH	$M/60$	10	1.1
Na₂CO₃	$M/60$	10	0.5
NaHCO₃	$M/60$	7	0.3
Na₂B₄O₇	$M/60$	6	0.8
Na₃PO₄	$M/60$	10	0.6
Na₂S₂O₃	$M/60$	5	0.8
KI + excess of KIO₃	$M/60$	5	1.2
Na₄(EDTA)	$M/120$	7	0.2
KIO₃ + excess of KI	$M/360$	6	0.2
NaH₂PO₂	(insufficient heat to give an end point)		
NaHSO₃	(insufficient heat to give an end point)		
\multicolumn{4}{c}{Titrations with M tetrasodium EDTA as titrant}
Ca(NO₃)₂	$M/60$	8	1.0
NiSO₄	$M/60$	11	0.6
CuSO₄	$M/60$	11	0.5
CdSO₄	$M/60$	11	0.5
ZnSO₄	$M/60$	5	0.4
BaCl₂	$M/60$	10	0.9
AgNO₃	$M/60$	4	2.0
CoCl₃	$M/60$	4	0.3
Ce(SO₄)₂	$M/120$	7	0.9
SnCl₄	$M/120$	10	0.6
Cr(NO₃)₃	$M/60$	3	0.2
AlCl₃	$M/60$	8	0.8
Mg(NO₃)₂	$M/60$	5	1.6
Be(NO₃)₂	$M/60$	6	1.4
LiNO₃	(insufficient heat to operate titrator)		
NH₄NO₃	(insufficient heat to operate titrator)		
Sr(NO₃)₂	$M/60$		
Hg(NO₃)₂	$M/60$		
Fe(NO₃)₃	$M/60$	(no results)	
Fe(NO₃)₂	$M/60$		
MnCl₂	$M/60$		

TABLE 3 (con't)

Titrate	C_m	No. of titrations	Sigma (%)[b]
Titration with M sodium carbonate as titrant			
Pb(NO$_3$)$_2$	$M/60$	4	1.4
AgNO$_3$	$M/30$	8	1.1
HCl	$M/30$	6	0.5
Zn(NO$_3$)$_2$	$M/60$	7	1.1
CuSO$_4$	$M/60$	6	0.8
AlCl$_3$	$M/90$	8	3.0

[a]Reprinted from [15, p. 201] by courtesy of the Society for Analytical Chemistry.
[b]Coefficient of variation of the scatter.

cations, Zr(IV), Th(IV), Cu(II), and UO$_2^{2+}$ [71]. However, U(IV), Fe(II), Fe(III), and Al(III) were found to interfere because of hydrolysis before the acid end point was reached, an example of a situation where the pK values were too close together. Phenols and cresols may be determined singly and in some cases in mixtures by titration with NaOH solution as shown by Paris and Vial [72]. Daftary and Haldar [73] used HCl and NaNO$_2$ solutions to determine alkyl amines, pyridine, nicotine, aniline, o-toluidine, and m-aminophenol singly and in mixtures. The incremental method of titrant addition allowed them to use the somewhat slow reaction of NaNO$_2$ with primary and secondary amines. The determination of OH$^-$ in Kraft black liquor by titration with phenol [74] and of the cation exchange capacity of clay suspensions by titration with sodium hydroxide [5] are further examples of the use of thermometric titration to analyze complex mixtures. The use of a weak acid titrant in one case (phenol) takes advantage of the method's ability to use a reaction with a small equilibrium

1. THERMOMETRIC TITRIMETRY   33

constant [see Eq. (1) and discussion thereof]. The determination of the number and type of prototropic groups on ovalbumin protein before and after diazotization with p-aminobenzoic acid is a related application of the method [75].

An example of a case in which a slow reaction rate is put to favorable use in the continuous method of titrant addition is the determination of Ca^{2+} as encountered in dolomite [$(Ca,Mg)CO_3$]. Jordan and Billingham [76] titrated mixtures of Mg^{2+} and Ca^{2+} with ammonium oxalate solution. They found the reaction with Mg^{2+} to be too slow to occur to any appreciable extent during the titration. The precipitation of calcium oxalate, however, was found to be rapid and to give a sharp end point on the thermogram, thus allowing the estimation of Ca^{2+} to be made in the presence of Mg^{2+}.

Rasmussen and Nielsen [77] used a KCN titrant to determine Ag(I), Ni(II), and Hg(II) by thermometric titrimetry. They found that Ag(I) could be titrated successfully in the presence of NH_3, but in the absence of NH_3 slow reactions (presumably hydrolysis of the metal ion) caused a fading end point. Ni(II) behaved similarly, but could be determined in an ammonia-ammonium chloride buffer. The end points observed corresponded to the formation of the dicyanoargentate(I) and tetracyanoniccolate(II) ions. A mixture of Ag(I) and Ni(II) failed to show any end points which could be used to determine either in the presence of the other. Under all conditions Hg(II) gave results which were ~1% too high and not as precise as those obtained for Ag(I) and Ni(II). In contrast to these results, Christensen et al. [78] found a sharp end point for formation of the tetracyanoniccolate(II) ion in the absence of any buffer and a stoichiometrically correct end point for formation of $Hg(CN)_2$ [42] (Fig. 9). The failure of Rasmussen and Nielsen to find these end points may have been due to poor mixing, addition of titrant too rapidly, or a combination of these effects.

Potassium iodide can be used as a titrant to determine Ag(I), Hg(II), or Pd(II) in very dilute solution by utilizing, to sharpen the end point, the reaction of Ce(IV) with As(III) as catalyzed by the addition of excess I$^-$ ion [48,50]. Thiocyanate can be used to back titrate these same metal ions [49]. By the addition of a known excess of any one of these metal ions to a solution containing any one of the following anions, Cl$^-$, Br$^-$, I$^-$, SCN$^-$, CN$^-$, Fe(CN)$_6^{4-}$, and S^{2-}, the anion concentration can be determined by back titration with 0.01 MKI titrant [50]. The iodide catalyzed end point indicator reaction was also used in these determinations.

Fluoride ion can be determined quickly and accurately in the presence of other anions by titration with a solution of Th(IV), Ce(III), Al(III), or Ca(II), the selection of titrant depending on which interfering anions are present [79]. Aluminum can be determined by the reverse titration with a known fluoride solution [80].

The precipitation reaction of potassium, ammonium, and thallium(I) ions with tetraphenylborate has been used to determine milligram quantities of these ions by thermometric titration [81].

B. Aqueous Redox Reactions

Potentiometric titration methods generally become inaccurate at concentrations below 0.1 M because the curve of voltage (or log [concentration]) versus milliliters of titrant becomes too flat for accurate determination of the end point. Thermometric titration methods, on the other hand, do not show this loss of end-point sharpness even in solutions more dilute

1. THERMOMETRIC TITRIMETRY 35

than 0.1 M [see Eq. (1) and discussion thereof]. The following examples will illustrate this point.

Izatt et al. [82] used potassium dichromate solution to determine Fe(II) in acid solution at concentrations of 0.05 M and 0.007 M. Their results are shown in Fig. 13 and Table 4. As can be seen in Table 4 the method gives results comparable to those obtained by conventional procedures and is not affected by a tenfold dilution of the Fe(II) solution.

Miller and Thomason [83] reported that the potentiometric titration of U(IV) with potassium dichromate was not feasible

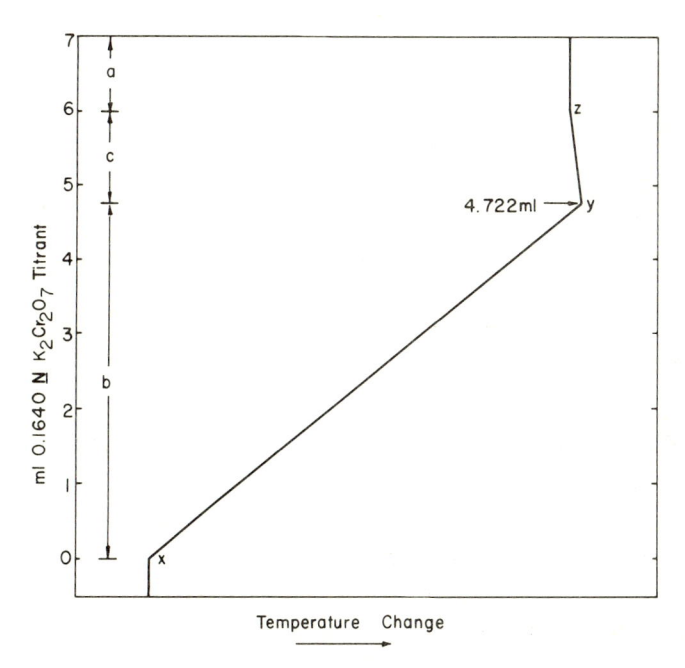

FIG. 13. THERMOGRAM FOR THE TITRATION OF 99.95 ML OF 0.007748 N FeSO$_4$ SOLUTION WITH 0.1640 N K$_2$Cr$_2$O$_7$: (a) INITIAL STIRRING SLOPE, (x) POINT AT WHICH TITRANT ADDITION BEGINS: (b) REACTION: $Cr_2O_7^{2-} + 6Fe^{2+} + 14H^+ = 2Cr^{3+} + 6Fe^{3+} + 7H_2O$, (y) EQUIVALENCE POINT: (c) CONTINUED ADDITION OF TITRANT, (z) TITRANT ADDITION TERMINATED: AND (d) FINAL STIRRING SLOPE.

TABLE 4

Standardization of Ferrous Sulfate Solutions
by a Thermometric Titration of Fe^{2+} With $K_2Cr_2O_7$[a]

$K_2Cr_2O_7$ titrant (ml)	$[Fe^{2+}]$ found, M	$[Fe^{2+}]$ added, M
0.9876 N $K_2Cr_2O_7$		
5.432	0.05332	
5.430	0.05330	
5.434	0.05334	0.05333
5.431	0.05331	
5.435	0.05335	
0.1640 N $K_2Cr_2O_7$		
4.724	0.007751	
4.722	0.007748	
4.721	0.007746	0.007746
4.719	0.007743	
4.719	0.007743	
4.722	0.007748	

[a] 99.95 ml solution of Fe^{2+}.

because of the poisoning and irreversibility of the Pt indicator electrode. However, using thermometric titration and the same reagents, they were successful in determining U(IV) at concentrations as low as ∼0.01 M.

Thiosulfate ion has been used in thermometric titration studies of several materials in the millimolar concentration range. Billingham and Reed [84] used thiosulfate ion to titrate the iodine liberated by the addition of Cu(II) to excess KI solution. They found good results with copper concentrations in the range 2-50 mM. Priestley [15] used thiosulfate ion as a titrant to determine the materials shown in

TABLE 5[a]

Thermometric Titrations With
2 M Sodium Thiosulfate as Titrant

Titrate	Molar concentration	No. of titrations	Coefficient of variation (%)
I_2 in KI	$M/60$	8	1.5
$Ce(SO_4)_2$ in H_2SO_4	$M/30$	8	0.6
Na_2SO_3 in KI_3 (titration of excess of I_2)[b]	$M/1000$	6	0.4
$NaHSO_3$ in KI_3 (titration of excess of I_2)[b]	$M/1000$	6	0.5
Hydroquinone in $K_2Cr_2O_7$ + H_2SO_4 (titration of excess of $Cr_2O_7^{2-}$)[c]	$M/4000$	4	1.4

[a] Reprinted from [15, p. 202] by courtesy of the Society for Analytical Chemistry.
[b] Maximum concentration $M/12$.
[c] Maximum concentration $M/50$.

Table 5. The accuracy obtained in his study as indicated in column 4 is remarkable in view of the extremely dilute solutions used in some cases.

Ceric sulfate solutions are useful for determining oxidizable substances by thermometric titration, especially in solutions which are generally too dilute for other methods. Results by Priestley [15] and Tyson et al. [12] using this titrant are given in Tables 6 and 7 respectively. Again, the results agree well with those determined by conventional procedures. Cerium(IV) has also been used as an oxidant to determine thiourea in mixtures with N-substituted thioureas [85].

TABLE 6[a]

Thermometric Titrations With $M/4$
Ceric Sulfate in Sulfuric Acid as Titrant

Titrate	Molar concentration	No. of titrations	Sigma (%)[b]
KI	$M/120$	8	1.5
Na$_2$SO$_3$	$M/240$	13	0.7
Na$_2$S$_2$O$_3$	$M/240$	8	0.4
Hydroquinone	$M/600$	8	0.4
Metol	$M/240$	7	0.4
Phenidone	$M/600$	11	1.8
Resorcinol	$M/480$	8	0.8

[a]Reprinted from [15, p. 202] by courtesy of the Society for Analytical Chemistry.
[b]Coefficient of variation of the scatter.

Two other oxidizing agents have also been used to determine organic functional groups. Sodium hypochlorite was used to titrate aromatic sulfonic acid amides [86] and sodium nitrite was used to determine amines which can undergo diazotization [73,87]. In both cases incremental titrant addition was used because of the slowness of the reactions. Accuracy of the results ranged from about 1 to 9%.

C. Reactions in Organic Solvents

An interesting use of thermometric titrimetry is the analysis of reactive organometallics in hydrocarbon solutions. For example, Everson [88] has determined butyllithium in hydrocarbons by titration with butanol, and Everson and Ramirez report the determination of diethyl zinc [89] and alkyl aluminum compounds [90] in hydrocarbon solutions. In the determination of diethyl zinc it was found that either o-phenanthroline or 8-quinolinol could be used to obtain a sharp

1. THERMOMETRIC TITRIMETRY

TABLE 7[a]

Results of Thermometric Titrations

Substance titrated	Titrant	Amount added (meq)[b]	Amount found (meq)[b]	Mean Deviation	Error (%)
H_2O_2	0.3477 N ceric sulfate in 3 M H_2SO_4	0.4114(3)	0.426$_7$	0.3	+3.7
$(-CH_2NH_3)_2SO_4$- $FeSO_4 \cdot 4H_2O$	0.3483 N ceric sulfate in 3 M H_2SO_4	0.2123(3)	0.211$_8$	0.4	-0.2
		0.356$_8$	0.351$_9$		-1.4
		0.339$_5$	0.337$_6$		-0.6
		0.350$_0$	0.350$_0$		0.0
		0.358$_7$	0.355$_0$		-1.0
		0.361$_2$	0.357$_3$		-1.1
		0.385$_3$	0.386$_4$		+0.3
		0.379$_0$	0.379$_1$		0.0
		0.349$_2$	0.350$_0$		+0.2
		0.179$_4$	0.178$_5$		-0.5
		0.189$_6$	0.192$_8$		+1.7
		0.180$_4$	0.182$_0$		+0.9
$K_4Fe(CN)_6 \cdot 3H_2O$	0.3483 N ceric sulfate in 3 M H_2SO_4	0.4510	0.443$_9$		-1.6
		0.4471	0.440$_0$		-1.5
		0.4554	0.448$_5$		-1.5

[a]Reprinted from [12, p. 1643] by courtesy of the American Chemical Society.
[b]Numbers in parentheses represent number of duplicate determinations performed.

end point corresponding to the formation of a 1:1 complex [89]. Everson and Ramirez [90] found for the reactions of alcohols, ethers, amines, ketones, and oxine with alkyl aluminum compounds, R_3Al, R_2AlH, R_2AlOR, and R_2AlCl, where R is a hydrocarbon chain (C_1-C_8) that all but R_2AlOR could be readily determined. Reactable impurities in the solvents used resulted in low precision unless a "clean up" titration was performed. This clean up consisted of adding some of the alkyl aluminum compound to be analyzed to the solvent in the reaction vessel, stirring for a period of time, titrating to the end point, and then proceeding with the actual analysis titration. Three or four more replicate titrations could then be run consecutively. If the end point was overrun in the clean up titration, the excess was simply included in the calculations on the following titration. Precise results were obtained by this procedure.

The above method for analysis of alkyl aluminum compounds was reversed by Crompton and Cope [91] in order to determine impurities in hydrocarbons. Triethyl aluminum dissolved in hydrocarbon was the titrant used to determine total reactable impurities such as water, oxygen, and alcohol in a hydrocarbon stream. Although the final procedure involved continuous flow enthalpimetry and not specifically thermometric titration, the initial development used the latter method.

Other reactions which involve the use of organometallic compounds are the titration of Grignard reagents in toluene with isopropyl alcohol [92] and the titration of organic acids and phenols in benzene, carbon tetrachloride, chlorobenzene, and dichloromethane with thallous ethoxide [93]. The titration of dioxane in benzene or carbon tetrachloride with stannic chloride titrant gives a sharp 1:1 end point and is an example of a Lewis acid-base reaction in these solvents [94]. Because they are usually fast, such reactions are potentially of more

1. THERMOMETRIC TITRIMETRY

interest for analysis of organic functional groups than the more usual organic reactions.

Quilty [95] has shown that the total acidity in petroleum products can be determined using a titrant consisting of potassium hydroxide dissolved in isopropyl alcohol. Snyder [96] has also described the use of an "onstream" thermometric titrator to determine acetic acid and sodium hydroxide in nonaqueous solvents.

As an example of the use of an inert solvent for protonic acid-base reactions, Mead [97] has studied the reactions of 35 different primary, secondary, and tertiary amines with trichloroacetic acid in benzene. Sharp 1:1 end points were found in all cases.

Acetonitrile has been used as a solvent for the titration of a wide range of acids and bases. Because of its nearly negligible acid-base properties, it exerts no leveling effect and hence has one of the widest ranges of acidities and basicities known for a solvent. (Very strong bases cannot be used, however, because of slow decomposition of the solvent in their presence.) At the same time, because of its polar nature, acetonitrile is a good solvent for a wide range of organic acids and bases. Forman and Hume [98] investigated the titration behavior of 32 amines and 13 carboxylic acids in acetonitrile. The titrants used for bases and acids were HBr and diphenylguanidine solutions, respectively, in acetonitrile. Table 8 gives representative data obtained by them. The determination of many of the substances successfully determined in acetonitrile is difficult in water because of low solubility and too small pK values.

Glacial acetic acid is also convenient to use and, because of its acidity as a solvent, it allows the titration of materials which are too weakly basic to be determined in water.

TABLE 8[a]

Representative Data for Thermometric Titration
of Some Amines and Acids in Acetonitrile Solvent

Compound	No. of determinations	Present (mmole)	Found (mmole)	Accuracy %	Standard deviation %
Titration of nonaromatic amines with HBr					
Di-*n*-butylamine	4	0.600	0.598	99.7	0.14
Diethylamine	3	0.595	0.598	100.5	0.40
Di-*sec*-butylamine	4	0.285	0.281	98.6	0.76
Di-isopropylamine	3	0.130	0.129	99.2	1.56
Tri-*n*-butylamine	4	0.598	0.595	99.5	0.56
Titration of aromatic amines with HBr					
N,N-Diethylaniline	4	0.457	0.449	99.1	1.03
p-Anisidine	3	0.640	0.634	99.1	0.40
Pyridine	4	0.590	0.584	99.0	0.74
p-Toluidine	4	0.828	0.824	99.5	0.23
N,N-Dimethylaniline	3	0.610	0.604	99.0	0.94
Titration of acids with 1,3-diphenylguanidine					
Benzoic	3	0.302	0.309	102.3	1.21
p-Chlorobenzoic	3	0.269	0.275	102.3	2.4
m-Chlorobenzoic	3	0.0910	0.0903	99.2	1.55
m-Bromobenzoic	3	0.240	0.243	101.3	1.35
m-Nitrobenzoic	3	0.449	0.445	99.1	1.05

[a] Reprinted from [98, pp. 132, 133, and 135], by courtesy of Pergamon Press, New York.

The analysis for water and acetic anhydride (which interfere with some determinations) in the glacial acetic acid solvent is conveniently done by thermometric methods [99]. Keily and Hume [13] have used perchloric acid dissolved in glacial acetic acid as a titrant for amines, salts of carboxylic and inorganic acids, some inorganic chlorides, acetamide, urea, and acetanilide. Some of the nitrogen bases studied, i.e, urea, acetamide, and acetanilide, are too weak to be titrated

1. THERMOMETRIC TITRIMETRY   43

potentiometrically even in this solvent, but good end points were obtained with thermometric titration. The carboxylic acid and inorganic acid salts were titrated directly while the chlorides were first metathesized to acetates by adding excess mercuric acetate.

Acetone is an interesting solvent for thermometric acid-base titrations because it acts as a thermochemical indicator on the addition of excess base [52]. The excess base reacts instantaneously with the acetone in what is probably a classical aldol condensation to give 4-hydroxy-4-methyl-2-pentanone. Carboxylic acids, phenols, β-diketones, imides, and CO_2 dissolved in acetone have been titrated with KOH in methanol [52]. The end point is indicated by a sharp change in the rate of temperature rise, as shown in Fig. 14. Acetone has also been

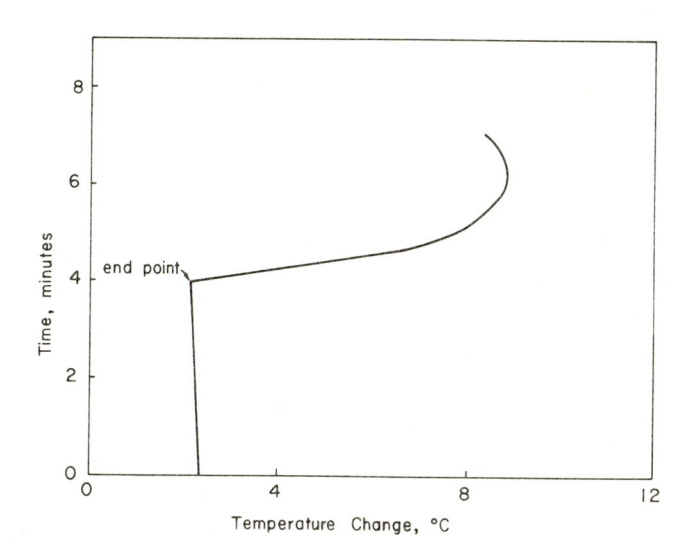

FIG. 14. THERMOGRAM FOR TITRATION OF AN ACID IN ACETONE WITH KOH IN METHANOL. REPRINTED FROM [52, p. 595], BY COURTESY OF THE SOCIETY FOR ANALYTICAL CHEMISTRY.

used as a solvent for the titration of phenolic groups in coal using KOH in isopropyl alcohol as the solvent [6].

Thermometric titration methods have also been used to determine the aprotonic acidity of solid oxide catalysts [100, 101]. Anhydrous suspensions in benzene of Al_2O_3 and aluminosilicate catalysts doped with various metals can be titrated with a Lewis base such as ethyl acetate.

Complexometric titrations of metal ions in nonaqueous solvents can be followed by thermometric methods. The titration of Ag(I) with chloride, bromide, or iodide in dimethylsulfoxide gives, in each case, two sharp breaks in the thermograms [102]. The breaks correspond to the stepwise formation of $AgCl_2^-$, $AgBr_2^-$, and $Ag_5I_7^{2-}$, respectively. Kiss [49] has shown that 16 different metal ions can be determined in dimethylsulfoxide after addition of known but excess amounts of EDTA, NTA, or diaminocyclohexanetetraacetic acid by back titration with Mn(II). Because the amount of heat liberated from reaction of the Mn(II) with the complexing agent was too small to give a good end point, a thermochemical indicator consisting of H_2O_2 and resorcinol was added to the titrate solution. Excess Mn(II) catalyzes the reaction of H_2O_2 and resorcinol and thus a sharp rise in temperature occurs at the end point. Amounts of metal as small as 1.5 µg were determined with a standard deviation of less than 1%.

Thermometric methods can also be used to detect the end points of coulometric titrations in nonaqueous solvents [103].

D. Reactions in Molten Salt Solvents

Jordan et al. [104-106] have titrated Cl^-, Br^-, I^-, and CrO_4^{2-} with Ag^+ at 158°C in a molten $LiNO_3$-KNO_3 eutectic solvent. The concentrations used were between 0.8 and 20 mM. Sharp end points were observed for all of the reactions

1. THERMOMETRIC TITRIMETRY

TABLE 9[a]

Determination of Potassium Chloride by Thermometric Titration in Fused Lithium-Potassium Nitrate[b] at 158°C

Initial chloride molality	Chloride, millimoles Taken	Chloride, millimoles Found	Experimental end point stoichiometry (moles) Ag^+/mole Cl^-	Relative titration error[c] (%)
8.615×10^{-4}	0.0939	0.0979	1.04	+4
2.112×10^{-3}	0.2133	0.2053	0.962	-3.8
5.205×10^{-3}	0.5231	0.5248	1.003	+0.3
1.170×10^{-2}	1.100	1.050	0.955	-4.5
1.985×10^{-2}	1.985	1.980	0.997	-0.3
mean values			0.991	-0.8
standard deviation of mean			±0.016	±1.5

[a] Reprinted from [105, p. 1440], by courtesy of the American Chemical Society.
[b] Eutectic melt, 34% (w/w) $LiNO_3$.
[c] Referred to theoretical equivalence point stoichiometry corresponding to AgCl.

corresponding to the precipitation of the silver salts. A summary of the results in the case of KCl is given in Table 9.

IV. DETERMINATION OF THE STOICHIOMETRY OF A REACTION

The stoichiometry of a reaction giving a sharp thermometric end point may easily be determined if the concentrations of the reactants are known. A sample thermogram (normalized by dividing the heat liberated by the number of moles reacted) is shown in Fig. 15 for the titration of Gd^{3+} with ethylenediamine in anhydrous acetonitrile [107]. Four distinct end points are seen corresponding to the formation of consecutive

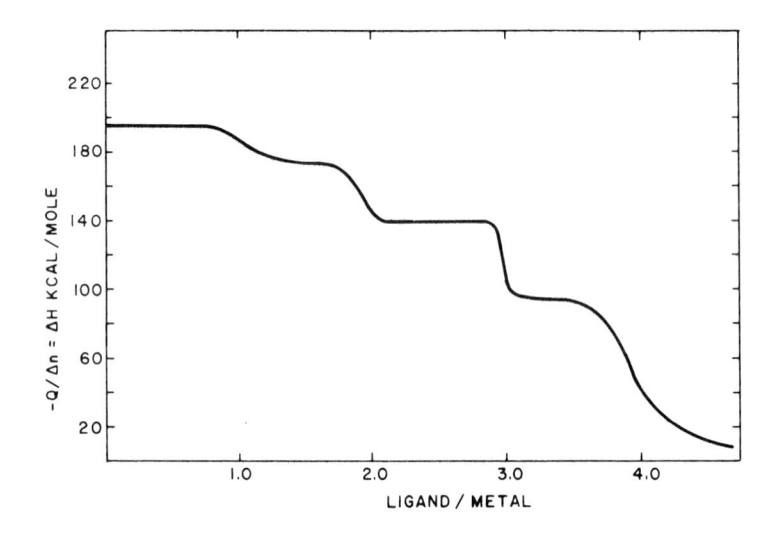

FIG. 15. THERMOGRAM FOR TITRATION OF Gd^{3+} WITH ETHYLENE-DIAMINE IN ANHYDROUS ACETONITRILE. REPRINTED FROM [107, p. 890], BY COURTESY OF THE AMERICAN CHEMICAL SOCIETY.

$Gd(en)_n^{3+}$ species. The particular applicability of this method to reactions in nonaqueous solvents is further shown by the determination of the stoichiometries of the reactions in benzene of Sn(IV) and Sb(V) chlorides with benzanthrone [108], the reactions in benzene of a variety of amines and ethers with $SnCl_4$ [109], and compositions of the species extracted into a variety of organic solvents [110].

Thermometric titrimetry can also be used to provide the necessary data for an independent check on the stoichiometry of an incomplete reaction since the ΔH value calculated for a correct reaction must be constant throughout the titration. Thus, if calculated values of ΔH vary with the total concentration of titrant or titrate, the assumed reaction is incorrect. Grenthe and Leden [111] used this principle to verify that Tl(III) chloride and bromide complexes were monomeric. They showed in a rather elegant fashion that the relation between $\bar{n}$ (the average number of bound ligands per metal ion) and Δh_V (the total molar heat change) was independent of the total concentration of metal ion. Arnek [112] has used the same principle (if not the same method) as an aid in determining the composition of the species formed during the hydrolysis of several metal ions. Schmulbach and Hart [113] demonstrated the existence of a 2:1 triethylamineiodine complex by showing that the enthalpy change calculated for the 1:1 reaction was a function of the total amine concentration.

A secondary, unexpected reaction between Ce^{4+} and α-mercaptocarboxylic acids was discovered by Alexander et al. [114] when they observed that the evolution of heat by the expected reaction (the formation of disulphide) was not constant throughout the titration.

Brenner [115] has treated the more general case of determining the formula of the species formed in the system. He has shown that for a pair of reactants, A and B, which form only one complex, A_mB_n, the ratio of m to n can be determined by measuring the partial molal heat effects of reactants A and B, λ_A and λ_B, respectively. λ_A is obtained by the addition of an increment of A to a solution of pure B. Since there is a large excess of B, A is quantitatively converted into the complex A_mB_n and λ_A is the heat of this reaction divided by the moles of A added to pure B. Similarly, an increment of B added to a large excess of A yields the maximum partial molal heat effect, λ_B. Since $m\lambda_A$ must equal $n\lambda_B$ (the heat of formation of A_mB_n is a constant) and since m and n must be integers, the ratio λ_A/λ_B establishes, as shown by Eq. (4), the empirical composition of the complex. This method has been used to determine the composition of the complexes formed in copper(I)-

$$\frac{m}{n} = \frac{\lambda_B}{\lambda_A} \quad (4)$$

cyanide solutions [115], in fused potassium chloride-cadmium chloride mixtures [116], and in fused sodium molybdate-molybdenum trioxide mixtures [117].

V. CALORIMETRIC EQUILIBRIUM CONSTANT DETERMINATION

A general overall treatment of the determination of equilibrium constants by titration calorimetry has been published in a series of three papers [118-120]. These three papers, Part I-- Introduction to Titration Calorimetry, Part II-- Data Reduction and Calculation Techniques, and Part III-- Application of Method to Several Chemical Systems, can be used to gain an understanding of the general usefulness of titration calorimetry for the determination of equilibrium constants.

A. Theory

The values of ΔG for a reaction in solution can be determined under certain conditions from a single thermometric titration thermogram [121]. In Fig. 16 a set of thermograms has been constructed (using corrected Q values and assuming ΔH for the reaction to be constant) showing the dependence of the shape of the thermogram on the equilibrium constant of the reaction occurring in the calorimeter. The successful application to a given system of the calorimetric titration method for determining equilibrium constants depends on (a) the equilibrium constant(s) and the reaction conditions being such that a measureable, but not quantitative, amount of reaction takes place ($-1 \leq \log K < 3$); and (b) the ΔH value(s) for the reaction(s) being measurably different from zero.

1. THERMOMETRIC TITRIMETRY

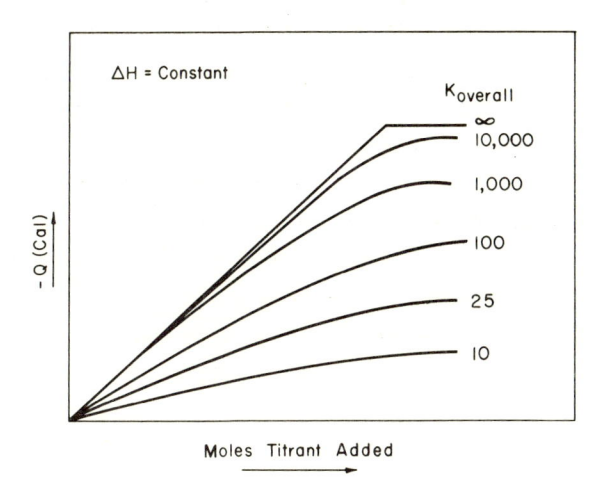

FIG. 16. PLOT OF CALCULATED Q(CAL) VALUES CORRECTED FOR ALL HEAT CHANGES EXCEPT THE CALORIMETER REACTION VERSUS MOLES OF TITRANT ADDED. TTHE Q(CAL) VALUES ARE CALCULATED FOR A SERIES OF K VALUES (RANGING FROM ∞ TO 10) ASSUMING IN EACH CASE A CONSTANT ΔH VALUE.

Consider the simple nonquantitative reaction of reactant A with B to form product AB (i.e., A + B = AB), which can be described by:

$$Q = \Delta H[AB]V \quad (5)$$

$$K = [AB]/[A][B] \quad (6)$$

$$[A_{total}] = [A] + [AB] \quad (7)$$

$$[B_{total}] = [B] + [AB] \quad (8)$$

where Q is the heat due to the chemical reaction, V is the volume of the solution, and [ ] indicates concentration of the bracketed species.

Equations (5)-(8) can be combined to give Eq. (9), which contains two unknowns, ΔH and K:

$$\Delta H/K = V[B_{total}][A_{total}]Q(\Delta H)^2 - [B_{total} + A_{total}]\Delta H + Q/V \quad (9)$$

Equation (9) may further be reduced to the general form of:

$$\Delta H/K = D_b(\Delta H)^2 + E_b\Delta H + F_b \quad (b = 1...N) \quad (10)$$

where the subscript b represents any one of the N data points obtained from the thermogram.

Equation (10) may be solved for values of ΔH and K using data from the thermogram for any two different values of b and obtaining a simultaneous solution of the two resulting equations [121]. However, it has been shown that such an approach can only give approximate values because the system is overdetermined [122]. The method of simultaneous solution apparently will not work at all on systems with multiple equilibria [123].

Alternatively, a least-squares method may be used. Equation (11) can be derived assuming that only random experimental errors appear in Q [84,85]:

$$U(K, \Delta H) = \sum_{b=1}^{N} W_b (Q_b^{obs} - Q_b^{calc})^2 \quad (11)$$

where W is a weighting factor and Q^{obs} and Q^{calc} are the observed and calculated heats, respectively. The correct or best K values are those which result in a minimum value for $U(K, \Delta H)$. The least-squares solution of $U(K, \Delta H)$ for K values is straightforward except that $\partial U(K, \Delta H)/\partial K$ is a nonlinear expression which cannot be solved directly and U_{min} must be found by some iterative method. A complete and accurate calculation of ΔH and K values thus involves five steps: (a) the assumption of initial values for the equilibrium constants; (b) the calculation of the concentration of each chemical species in the reaction vessel at each data point using the assumed K values; (c) the least-squares calculation from the expression $\partial U(K, \Delta H)/\partial \Delta H$ of the ΔH values corresponding to the assumed K values; (d) the calculation of the value of $U(K, \Delta H)$;

1. THERMOMETRIC TITRIMETRY

and (e) the selection of other values for the equilibrium constants and reiterating steps (b), (c), and (d) until the minimum value of $U(K, \Delta H)$ is found.

For reactions involving one equilibrium constant, step (a) [121] may be carried out by applying the method of simultaneous solutions to Eq. (10) in order to find an initial value for K.

Step (b) is carried out by use of the equilibrium constant expressions, mass balance equations, and charge balance equations. Combination of these equations results in an equation containing only one unknown concentration. This equation, which contains only one real, positive root (Descartes' rule of signs), may be easily and exactly solved by Newton's method of successive approximation.

Step (c) should be carried out by a least-squares fit of the Q values and the data obtained in step (b). Substituting

$$Q_b^{calc} = \Delta n_1 \Delta H_1 + \Delta n_2 \Delta H_2 + \ldots = \sum_1^p \Delta n_{p,b} \Delta H_p \quad (12)$$

into Eq. (11) gives

$$U(K_p, \Delta H_p) = \sum_{b=1}^N W_b (Q_b^{obs} - \sum_1^p \Delta n_{p,b} \Delta H_p)^2 \quad (13)$$

where the subscript b is over all data points and the subscript p is over all reactions being studied. The best ΔH_p values for the K values assumed can be calculated by means of a least-squares analysis of p equations of the form of Eq. (14) which results from setting the expression $\partial U(K_p, \Delta H_p)/\partial \Delta H_p$ from Eq. (13) equal to zero.

$$\sum_{b=1}^N (W_b Q_b^{obs} \Delta n_{k,b}) = \sum_{b=1}^N (W_b \Delta n_{k,b} \sum_I^p (\Delta n_{p,b} \Delta H_p)) \quad (14)$$

$$(k = 1 \ldots p)$$

Step (d), the calculation of the numerical value of $U(K_p, \Delta H_p)$, is accomplished by calculating Q_b^{calc} values from Eq. (12) and inserting these into Eq. (11).

Step (e), the reiteration of steps (b), (c), and (d) with selected K values to locate the minimum value $U(K_p, \Delta H_p)$, can be done in one of three ways [123]: (a) schematic mapping, (b) pit mapping, and (c) variable metric minimization. Schematic mapping consists of varying K over a large domain and finding the associated minimum value of $U(K, \Delta H)$ by trial and error. This method requires the longest calculation time of the three methods. Pit mapping as described by Sillen [124] and used by Paoletti et al. [125] and Wanders and Zwietering [122] consists of assuming a functional relation for $U(K, \Delta H)$ and evaluating U_{min} by direct differentiation, yielding new K and ΔH values which may be used in the next iteration. The difficulties which may be encountered in using pit mapping are that the area of convergence is small and that saddlepoints as well as minima will be found. Variable metric minimization as developed by Davidon [126] and used by Izatt et al. [123] differs from pit mapping in that the actual value of the derivative of $U(K, \Delta H)$ is used. This method eliminates the difficulties present in pit mapping, but the calculations are somewhat more complex.

If the Q_b^{obs} data are essentially free from systematic or random errors the minimum value of $U(K_p, \Delta H_p)$ should be close to the value of U_{min} estimated from the precision of the calorimetric equipment:

$$U_{min} = N\delta Q^2 \qquad (15)$$

where δQ is the estimated precision of the heat measurement at each point. The weighting factor W_b which appears in Eq. (11)-(14) has been set equal to 1.0 in all previous work. A suggested function for W_b which would weight each of the measurements according to the relative error in the measurement is

$$W_b = 1 - (\delta Q/|Q_b^{obs}|) \qquad (16)$$

Equation (16) becomes meaningless, however, for values of $|Q_b^{obs}|$ less than δQ, and should not be used in those cases.

The successful application of the calorimetric titration method for equilibrium constant determination requires the accurate measurement of the heat liberated in the titration. An exact method of analyzing thermometric titration curves for heat data is given in [121],[119] and [45].

B. Chemical Systems Studied by the Calorimetric Titration Method

The calorimetric titration method has been used to determine equilibrium constants for many different chemical systems such as proton ionization, metal ligand interaction, and adduct formation in aqueous and nonaqueous solutions.

In Table 10 pK values determined by the calorimetric titration method for several representative acids are given and are compared with values determined by other methods where such values are available. The pK values above four were determined by a competitive reaction scheme to be discussed later. It can be seen from the data given in Table 10 that this method is ideally suited to studying proton ionization from very weak and strong acids in aqueous solution. This is in contrast to most of the conventional methods of measuring pK values which become inaccurate when applied to weak or strong acids.

The application of the calorimetric titration method to pK regions difficult to study by other methods has resulted in at least one significant discovery, namely, a knowledge of the site of proton ionization and the reason for the acidic character of adenosine(II) [127]. A calorimetric titration study of adenosine and related compounds showed that both the 2'- and 3'-hydroxyl groups on the ribose moiety must be present for measurable acidic character to exist in aqueous solution.

TABLE 10

pK Values Determined by
Calorimetric Titration and by Other Methods

Acid	pK value determined by calorimetric titration[a]	pK value determined by other methods
Methyl-o-aminobenzoic	2.36 ± 0.03 [128]	2.32[128], 2.33[129], 2.24[130]
o-Aminobenzoic (pK$_1$)	2.09 ± 0.02 [128]	2.17[131], 2.05[129], 2.11[132], 2.14[133], 2.05[134], 2.14[135]
Methyl-m-aminobenzoic	3.58 ± 0.02 [128]	3.61[128], 3.55[136], 3.55[137], 3.56[130]
m-Aminobenzoic (pK$_1$)	3.07 ± 0.05 [128]	3.07[128], 2.90[129], 3.08[138], 3.12[132], 3.07[134], 3.05[137], 3.09[135]
Methyl-p-aminobenzoic	2.45 ± 0.02 [128]	2.50[128], 2.46[139], 2.40[134], 2.38[140]
p-Aminobenzoic (pK$_1$)	2.41 ± 0.04 [128]	2.43[128], 2.33[129], 2.45[139], 2.41[132], 2.49[141], 2.37[135], 2.38[134]
1,2,3-Triazole-4,5-dicarboxylic (pK$_1$)	1.94 ± 0.04 [142]	1.86[142]
Hydrogen sulfate ion	1.97 ± 0.03 [44]	1.91-1.99[44,121]
Monohydrogen phosphate ion	12.38 ± 0.03 [143]	12.37[144]
Ribose	12.22 ± 0.04 [47]	
Fructose	12.27 ± 0.01 [47]	
Adenosine	12.35 ± 0.03 [47]	
Adenosine 5'-monophosphate	13.06 ± 0.10 [47]	

[a]Uncertainties are given as standard deviations.

By use of a competitive reaction scheme, the calorimetric titration technique has been extended to acids with pK values between 3 and 11. For example, the pK value for proton ionization from protonated pyridine, reaction (17), is easily obtained by combining the pK$_3$ value for reaction (19) which

1. THERMOMETRIC TITRIMETRY

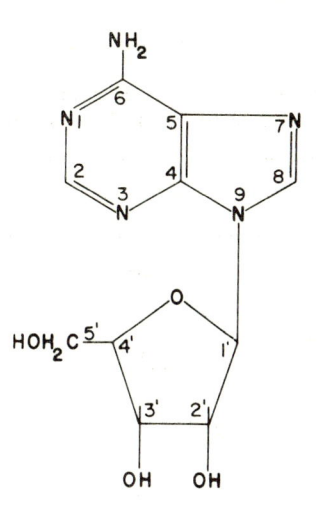

(II)

can be calculated from calorimetric titration data with the pK value for reaction (18) which is known from previous experiment or obtained from the literature. The $\Delta H°$ values in reactions (17), (18), and (19) are given in kilocalories per/mole.

$C_5H_5NH^+ = H^+ + C_5H_5N$ $pK_1 = 5.17$, $\Delta H° = 4.98$ (17)

$CH_3COOH = H^+ + CH_3COO^-$ $pK_2 = 4.76$, $\Delta H° = -0.01$ (18)

$C_5H_5NH^+ + CH_3COO^- = C_5H_5N + CH_3COOH$ $pK_3 = 0.41$,
$\Delta H° = 4.99$ (19)

This method has been applied to the determination of pK values for several substances including metanilic acid (HMet, pK = 3.756), protonated pyridine (HPyr$^+$, pK = 5.17), protonated imidazole (HIm$^+$, pK = 6.986), protonated THAM (HTHAM$^+$, pK = 8.069), and glycine (HGly, pK = 9.780) using acetic acid (pK = 4.7650) as a titrant. The results as given in Table 11 for the first four acids listed, are in excellent agreement with pK values determined by other methods. In the case of glycine where the reaction in the calorimeter has a pK value of 5.015, the pK value is seen to be in only fair agreement with other values. The calorimetric titration method can thus be extended

TABLE 11

Comparison of pK Values Determined by Calorimetric
Titration Using Acetic Acid as a Titrant and by Other Methods

pK	Method[a]	Ref.
\multicolumn{3}{c}{HMet = Met⁻ + H⁺}		
3.808	I	43
3.756	II	43
3.738	III	145
\multicolumn{3}{c}{HPyr⁺ = Pyr + H⁺}		
5.168	I	43
5.18	IV	146
5.17	(not stated)	147
\multicolumn{3}{c}{HIm⁺ = Im + H⁺}		
6.986	I	43
6.953	III	148
6.993	III	149
\multicolumn{3}{c}{HTHAM⁺ = THAM + H⁺}		
8.030	I	43
8.076	III	150
8.075	III	151
8.069	III	152
\multicolumn{3}{c}{HGly = Gly⁻ + H⁺}		
9.59	I	43
9.778	III	153
9.780	III	154
9.778	III	155

[a]Method I: Calorimetric titration (acetic acid titrant, ligand concentration = 0.010 M). Method II: Calorimetric titration (acetic acid titrant, ligand concentration = 0.016 M). Method III: Hydrogen electrode (potentiometric measurements). Method IV: Glass electrode (pH measurements).

1. THERMOMETRIC TITRIMETRY

to include any acid-base system provided a suitable titrant can be found. A complete set of titrants for studying proton ionization reactions in aqueous solution over the pK range 15 to -1 has been suggested by Christensen et al. [156]. Several other workers have also determined equilibrium constants by calorimetric procedures. Raffa et al. [157] determined $\Delta G°$ for the acid dissociations of ephedrinium and pseudoephedrinium ions by the calorimetric titration method and found excellent agreement with potentiometric values previously determined by other workers. Gill and Farquhar [158] used a variation of the calorimetric titration method based on heats of dilution to determine the equilibrium constants for self association of urea and purine and found agreement with values determined by other methods. Epley and Drago [159] have used incremental calorimetric titrations to determine equilibrium constants for adduct formation. Their results on the reaction between phenol and a series of Lewis bases in carbon tetrachloride or cyclohexane solution and on the reaction between $(CH_3)_3SnCl$ and N,N-dimethylacetamide are given in Table 12. They concluded that the K and ΔH values obtained from calorimetric data were more reliable than those obtained from spectrophotometric data. Neerinck et al. [160] have used a variation on the method used by Epley and Drago [159] to determine $\Delta G°$ and $\Delta H°$ values for the interaction of a series of phenols with tri-n-butylamine. Olofsson [161] has used the calorimetric titration method to obtain $\Delta G°$ and $\Delta H°$ values for the interaction of antimony pentachloride with water, substituted ethyl acetates, and various esters all in ethylene chloride solution. As a final example of the broad range of chemical systems that have been studied by this method, Benzinger [162] has determined the equilibrium constant for the hydrolysis of adenosine triphosphate to adenosine diphosphate and phosphate as catalyzed by ATPase.

TABLE 12

Equilibrium Constant Values
Determined by Calorimetric Titration

Base	K, 1/mole[a]
Phenol-base adducts in CCl_4[159]	
CH_3CN	4.8 ± 0.2
$CH_3COOC_2H_5$	7.2 ± 0.4
CH_3COCH_3	11.9 ± 1
C_4H_8O	13.5 ± 0.5
$CH_3CON(CH_3)_2$	107 ± 15
Phenol-base adducts in cyclohexane[159]	
C_5H_5N	79 ± 10
$(C_2H_5)_3N$	90 ± 10
Trimethyltin chloride-base adduct in CCl_4[172]	
$CH_3C(O)N(CH_3)_2$	3.4

[a]For meaning of uncertainties, see indicated references.

The calorimetric titration method has also been extended to the determination of log K values for weakly interacting metal-ligand systems involving one or more complexes. Examples are the Ag^+-pyridine (see Fig. 17 for a typical thermogram and plot of the species distribution) [123,125] and Cu^{2+}-pyridine [123] systems where $Agpy_2^+$ and $CuPy_4^{2+}$, respectively, are formed by consecutive addition of Py to the metal ions. The equilibrium constants obtained for these systems in water are given in Table 13. Paoletti et al. [163] have also studied the silver(I)-pyridine system in 50% water-dioxane as a solvent.

1. THERMOMETRIC TITRIMETRY

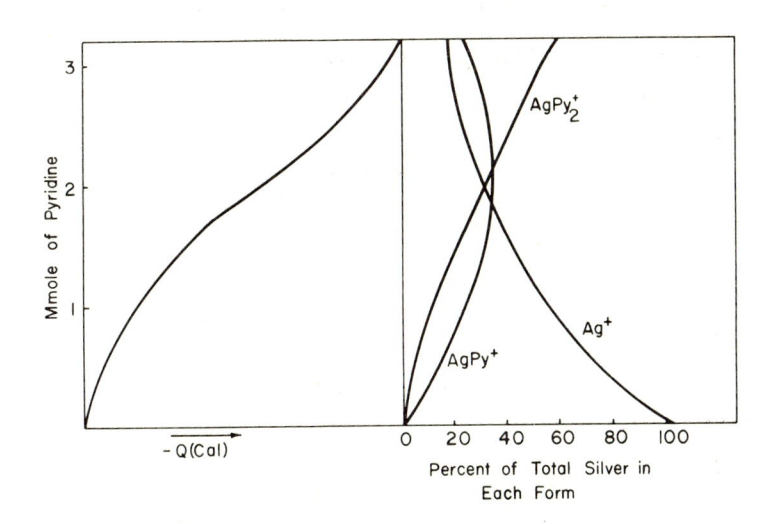

FIG. 17. THERMOGRAM AND SPECIES DISTRIBUTION FOR TITRATION OF Ag(I) WITH PYRIDINE. REPRINTED FROM [123, p. 1212] BY COURTESY OF THE AMERICAN CHEMICAL SOCIETY.

It is interesting that 12 thermodynamic quantities (four sets of ΔG, ΔH, and ΔS values) can be determined for the Cu^{2+}-Py system from a single thermometric titration. Becker and Luschow [164] have also successfully studied a four-step equilibrium system. They report pK, ΔH°, and ΔS° values for formation of the four metal-ligand complexes formed by the consecutive reaction of $4I^-$ with Cd^{2+}. Equilibrium constants for the interaction of SO_4^{2-} with Na^+, K^+, Mg^{2+}, Ca^{2+}, Mn^{2+}, Fe^{2+}, Co^{2+}, Ni^{2+}, Zn^{2+}, Cd^{2+}, Al^{3+}, Ga^{3+}, In^{3+}, Sc^{3+}, Y^{3+}, La^{3+}, and all the trivalent rare earth metal ions except Pm^{3+} have also been determined by this technique [44,165,166]. Another interesting application of the method is the determination by Harris and Moore [167] of the equilibrium constants

TABLE 13

Log β_i Values for the Ag^+-Pyridine and Cu^{2+}-Pyridine Systems Calculated From Calorimetric Titration Data. VAlues Valid in Aqueous Solution at 25°C

i	Log β_i
$Ag^+ + i\ Py = Ag(Py)_i^+$	
1	2.00[a]
1	2.05[b]
2	4.1[a]
2	4.10[b]
$Cu^{2+} + i\ Py = Cu(Py)_i^{2+}$	
1	2.50 ± 0.02[b]
2	4.30 ± 0.05[b]
3	5.16 ± 0.06[b]
4	6.04 ± 0.10[b]

[a]Ref. [125].
[b]Ref. [123]. Uncertainties are given as standard deviations.

for the five simultaneous equilibria involved in the titration with water of solutions of cobalt perchlorate and nickel perchlorate in butanol. The determination of equilibrium constants for the interaction of K^+, Rb^+, Cs^+, NH_4^+, Ag^+, Sr^{2+}, and Ba^{2+} with the cyclic polyether, dicyclohexyl-18-crown-6 [168], the interaction of sodium dodecylsulfate with cetylpyridinium chloride [169], and the adsorption of halide ions on silver citrate sols [170] are other examples of the application of the calorimetric titration method to systems that are very difficult to study by other means. The examples given above show the calorimetric titration method to be useful for the rapid, accurate determination of equilibrium constants in both aqueous and nonaqueous solvents.

1. THERMOMETRIC TITRIMETRY   61

By titrating the protonated form of a ligand into a solution of metal ion, the proton competes with the metal ion for the ligand, thus reducing the equilibrium constant for the reaction which occurs in the calorimeter to a small value. The calorimetric titration procedure for obtaining equilibrium constants thus becomes applicable to the study of very stable complexes. Eatough [171] applied this variation of the method to the determination of the equilibrium constants for the consecutive reaction of two 2-aminoethanol molecules with Hg^{2+}, and of three 1,10-phenanthroline molecules with Cu^{2+} and Zn^{2+}. The equilibrium constants determined by this method are in good agreement with values determined for these three systems by other methods.

The methods used to calculate equilibrium constants from calorimetric titration data vary widely and several authors have discussed the problems involved in their calculation [115, 121-123,125,164,167,172,173,174,175,176]. Three recent papers have attempted to deal with the problem of errors in the determination of K values, and, curiously enough, they arrive at quite different conclusions concerning the importance of errors propagated in the calculation methods [156,177,178]. Cabani and Gianni [177] concluded that equilibrium constants determined by this method are prone to large errors and should be checked by an independent method and that the method is invalid for systems with more than one chemical reaction. From a very similar treatment, Christensen et al. [156,178] concluded that with the equipment presently available, equilibrium constants could be determined by the calorimetric titration method as accurately as with other conventional methods (pH titration, spectrophometric, etc.) under the proper circumstances. This difference of opinion probably results from Cabani and Gianni assuming quite large systematic errors and not examining the effect of the magnitude of ΔH on the calculated pK and ΔH values. However, it is quite clear that for each particular application of the method a careful assessment of the errors present both in the data and in the calculations should be made [122,172].

VI. EQUIPMENT

A. Design and Construction of Thermometric Titration Equipment

Reported thermometric titration equipment varies from simple to very elaborate in design, the degree of sophistication depending on the accuracy, sensitivity, and kind of data required. In the case of a continuous titrant delivery system, this choice is largely dictated by the criterion that the temperature recorded at any time must be the "true temperature" of the solution being titrated. Because the variables being recorded in a continuous thermometric titration are temperature and volume of titrant (time), there must be no appreciable thermal lag in the system. Two approaches which can be used to achieve this condition are: (a) the titration time can be made short enough so that no appreciable heat loss occurs from the system during the titration; and (b) the reaction vessel can be made with a very small heat capacity and well-defined boundaries so that the system achieves nearly instantaneous thermal equilibration. Because the reaction vessel can be as simple as an insulated beaker, the first approach results in simple, inexpensive equipment suitable for analytical work of moderate precision (1-5%) for reactions having moderate to large enthalpy changes. The precision is limited largely because of the short time and rapid titrant delivery rate involved. The second approach involves much more elaborate equipment, but has the advantages that (a) analytical work of high precision (0.1-1%) can be accomplished; (b) reactions with small enthalpy changes can be used for analytical determinations; (c) long time periods and slow titrant delivery rates can be used if needed; and (d) quantitative calorimetric data may be obtained. The incremental method of titrant addition does not have as stringent requirements as those stated above for the continuous method, but does have the same design criteria as those used in constructing a conventional calorimeter.

B. Description of Commercially Available Titration Equipment

The first commercially available thermometric titrator was the Titra-Thermo-Mat (Aminco) based on a design by Jordan [25]. This unit consists of a 30-ml beaker in an insulated enclosure, a motor-driven syringe, a constant speed stirrer, and a Wheatstone bridge with a thermistor. The syringe delivers titrant into the beaker at a rate of 0.6 ml/min (Fig. 18).

The unit is capable of performing end-point determinations and other analytical operations, but not equilibrium constant determinations, with moderate precision (1-5%). In 1970 the unit was modified considerably and although the new unit is much easier to operate, the other characteristics are much the same. The newer unit has an 8-ml reaction vessel so that smaller samples can be used, and a variable speed buret arrangement based on a Sage pump.

A series of higher precision instruments based on the designs of Christensen et al. [179,180] is presently marketed by Tronac (Orem, Utah). The units were designed primarily as calorimeters, but are also excellent high precision (0.1-1%) analytical devices. In general the units have the following design characteristics. The reaction vessel consists of a

a

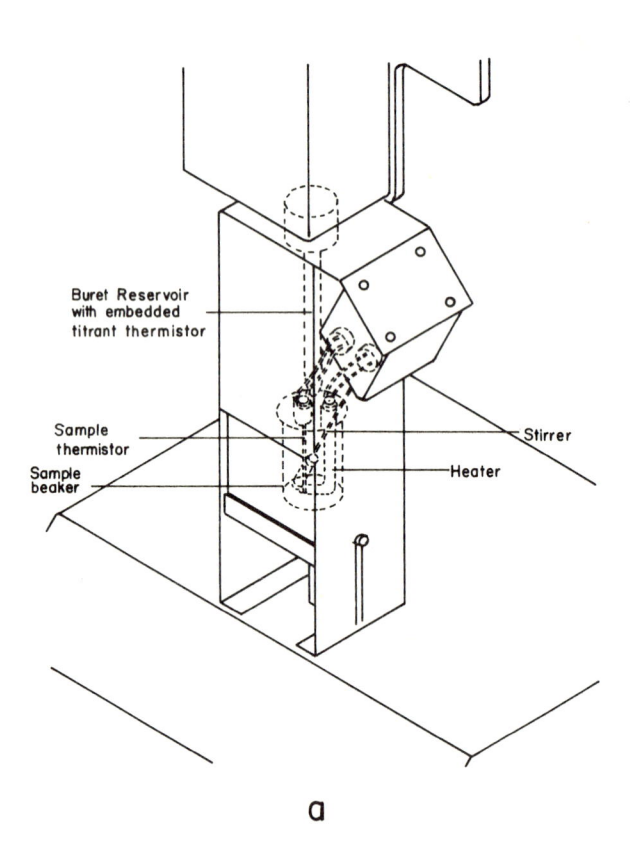

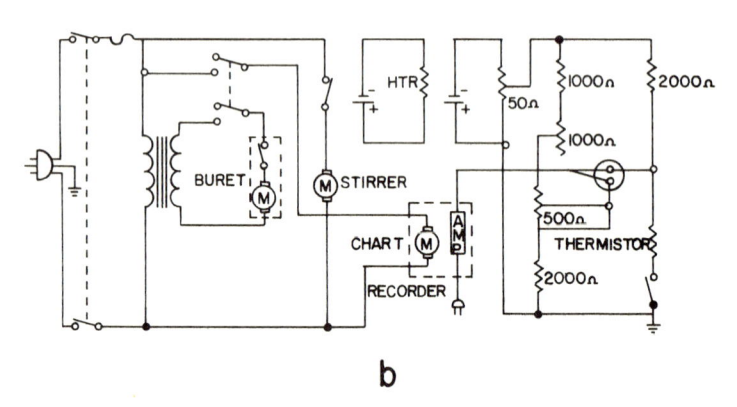

b

glass or metal container constructed in the shape of a round-bottom flask in order to reduce the heat leak path and the amount of material forming the boundary between the calorimeter and the surroundings. Both the calorimeter and titrant are submerged in a constant temperature (±0.0003°C) water bath. Titrant is delivered from a buret constructed from a Gilmont micrometer syringe which has a total capacity of 0.25, 1.0, 2.5, or 10.0 ml with quick change parts. The buret is driven by either a synchronous gear motor or a stepping motor. The titrant delivery rate is variable over a wide range and the buret may also be used in an incremental mode. The temperature measuring circuit is a dc Wheatstone bridge with either a 7.5KΩ or 100KΩ teflon encased thermistor (depending on output device) as one leg. The heater circuit is a conventional circuit with a 100-Ω teflon encased heater. The initial thermal equilibration is accomplished automatically by a transistor circuit which uses the thermistor and calibration heater. The various components of the Tronac solution calorimeters are shown in Figs. 19-22. The calorimeters are also available as isothermal models in which the data appear in units of μcalories/second rather than temperature. Inexpensive, much less automated bench models are also available. The Model 550 is shown in Fig. 23.

FIG. 18. TITRA-THERMO-MAT THERMOMETRIC TITRATOR. (a) "ADIABATIC TOWER" WITH MICROCALORIMETRIC CELL IN "TITRATION POSITION." STIRRER CONNECTED VIA FLEXIBLE SHAFT TO SYNCHRONOUS MOTOR. (b) ELECTRICAL CIRCUITRY. M: CONSTANT SPEED MOTORS. BURET: MENISCO-MATIC UNIT (AMERICAN INSTRUMENT CO., SILVER SPRING, MD.) USED IN TITRATIONS. HTR: ELECTRIC HEATER USED TO ADJUST SAMPLE TEMPERATURE OR CALIBRATE THERMISTOR BRIDGE RESPONSE, OR BOTH. RECORDER: 1 mV dc POTENTIOMETER USED TO MONITOR UNBALANCE POTENTIAL OF THERMISTOR BRIDGE. THERMISTOR: "TITRANT OR SAMPLE THERMISTOR" CONNECTED ALTERNATIVELY, AS DESIRED.

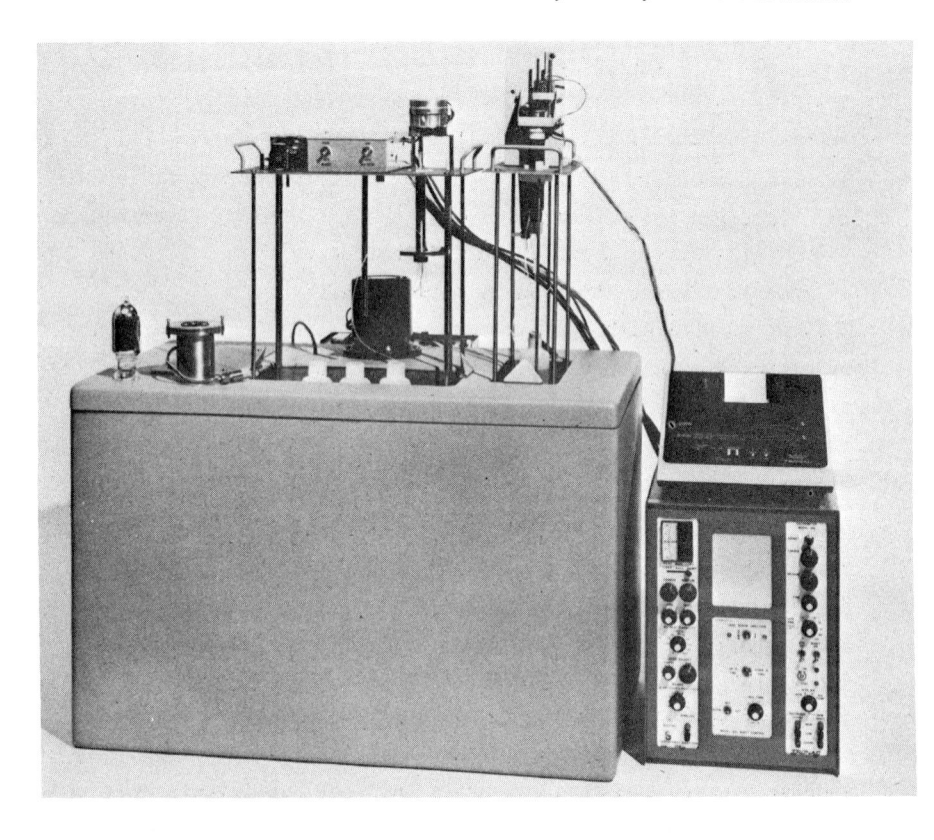

FIG. 19. TRONAC'S ISOPERIBOL AND ISOTHERMAL CALORIMETER. PUBLISHED BY COURTESY OF TRONAC INC., PROVO, UTAH.

A third commercial unit, which may be used for incremental titrations, is marketed by LKB Instruments, Inc. "The LKB calorimeter consists of a thin-walled pyrex-glass reaction vessel fitted with a 2,000 ohm thermistor, a 50 ohm calibration heater, and an 18-carat gold stirrer, the whole being contained in a chromium plated brass case. The case is submerged in a thermostated water bath in which the bath temperature is sensed

1. THERMOMETRIC TITRIMETRY 67

1. Bridge reference resistor tube
2. Header for reaction vessel
3. Isoperibol Dewar
4. Isothermal reaction vessel

INSERT ASSEMBLY

1. Stepping motor
2. Gilmont micrometer syringe
3. Reservoir

MODEL 822 PRECISION TITRATOR

FIG. 20. TRONAC CALORIMETER INSERTS. PUBLISHED BY COURTESY OF TRONAC INC., PROVO, UTAH.

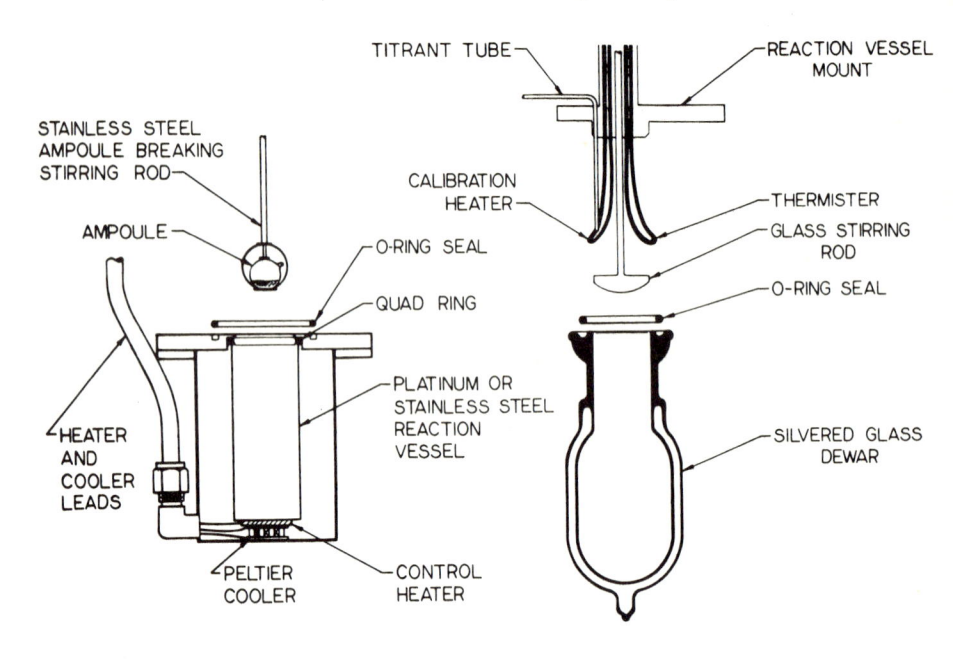

FIG. 21. METAL ISOTHERMAL REACTION VESSEL AND ISOPERIBOL GLASS DEWAR. A GLASS STIRRING ROD OR AN AMPOULE HOLDING STIRING ROD CAN BE USED ON EITHER REACTION VESSEL. PUBLISHED BY COURTESY OF TRONAC INC., PROVO, UTAH.

by a thermistor and regulated by proportional heating, providing, under normal operating conditions, a temperature constancy better than 0.001°. The space between the reaction vessel and the brass case may be evacuated in order to decrease the heat exchange." (Reprinted from [181, p. 3] by courtesy of LKB Instruments, Inc.) The details of the equipment are shown in Figs. 24-26.

C. Construction of a Simple Inexpensive Thermometric Titrator

The following discussion is based on the authors' experience with a number of thermometric titrators of different

1. THERMOMETRIC TITRIMETRY   69

MODEL 808 PROGRAMMER

Digital Display and Record

Automatic Control of Experiment

550 ISOTHERMAL   450 IOSPERIBOL CONTROL

    METER DISPLAY

                                                                                             DC WHEATSTONE BRIDGE

AC WHEATSTONE BRIDGE

                                                                                             CALIBRATION HEATER

PELTIER COOLING CONTROL

μ CALORIES PER COUNT TO
    CONTROL HEATER
              OUTPUT   DVM and RECORDER
                                                                                     OUTPUTS

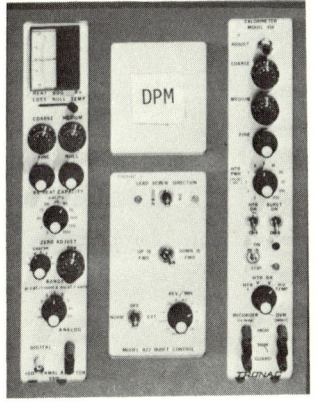

MODEL 822 SPEED CONTROL

FIG. 22. TRONAC'S ELECTRONIC CONTROL CONSOLES. PUBLISHED
BY COURTESY OF TRONAC INC., PROVO, UTAH.

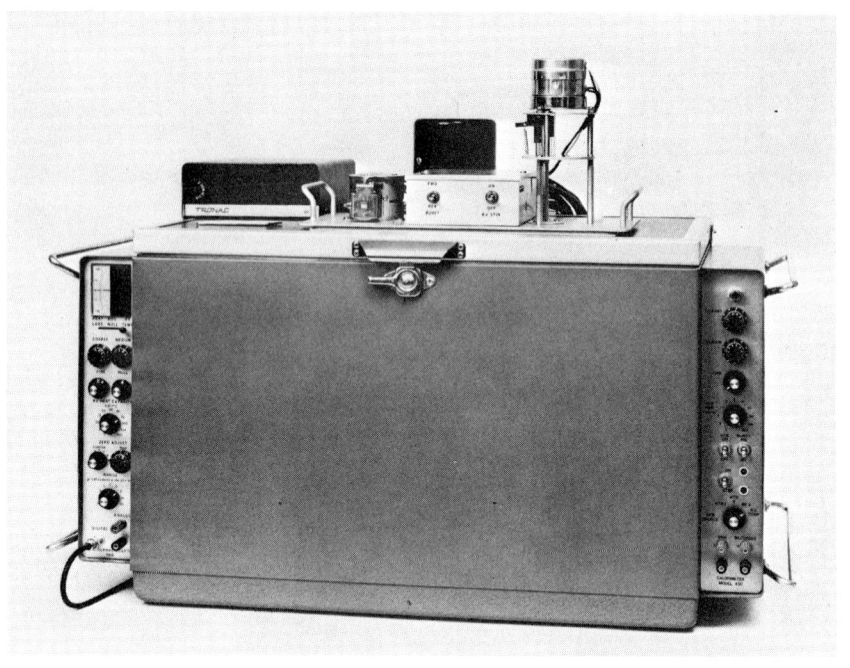

FIG. 23. OVERALL VIEW OF TRONAC 500 MODEL THERMOMETRIC TITRATOR AND CALORIMETER. PUBLISHED BY COURTESY OF TRONAC INC., PROVO, UTAH.

design, and the availibility and cost of components. The equipment described in the following pages can be built quickly and inexpensively and is suitable for many analytical applications as well as teaching [182]. The precision is in the 1-5% range and the temperature sensitivity is good enough to allow the analytical determination of materials in dilute solution. It should be clearly pointed out, however, that the equipment described in the following discussion is not suitable for accurate determination of heats of reaction or equilibrium constants.

The temperature sensing circuit is shown in Fig. 27. The thermistor has a temperature coefficient of -4.4%/°C so the

1. THERMOMETRIC TITRIMETRY

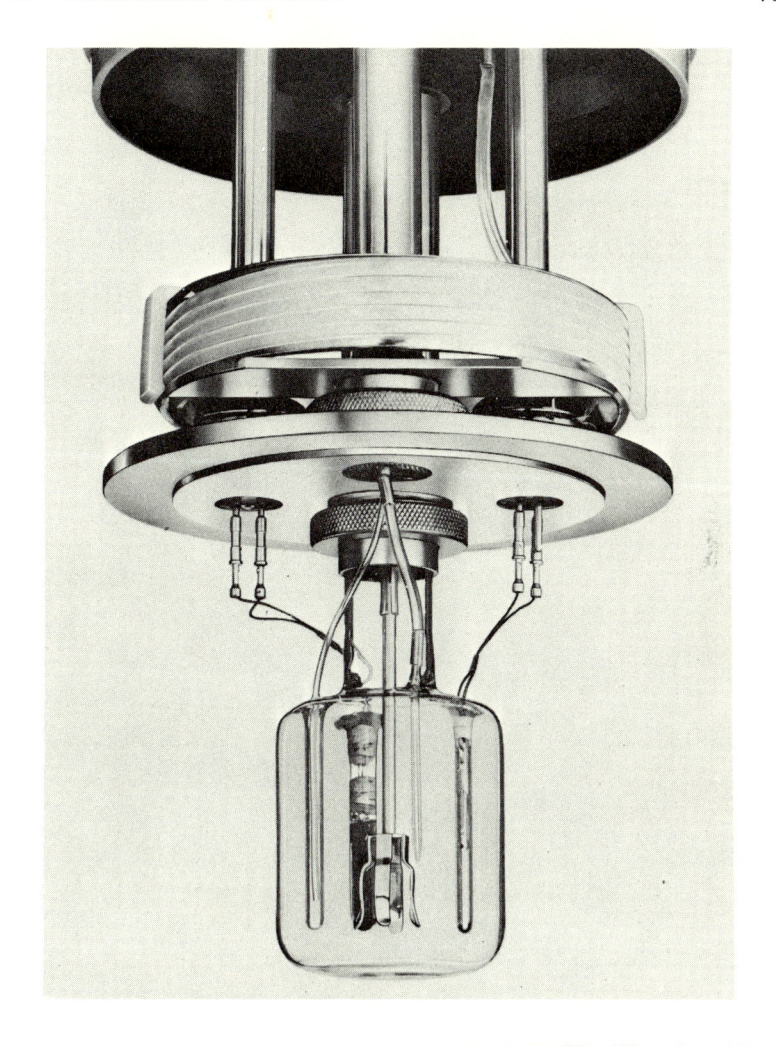

FIG. 24. DETAILED VIEW OF LKB CALORIMETER VESSEL. RE-PRINTED FROM [181] BY COURTESY OF LKB INSTRUMENTS, INC.

bridge sensitivity is ∼30 mV/°C. A 10-KΩ bridge was chosen because the resistance is large enough to give a high temperature sensitivity and yet small enough so that impedance matching problems will not be encountered with most recorders. A

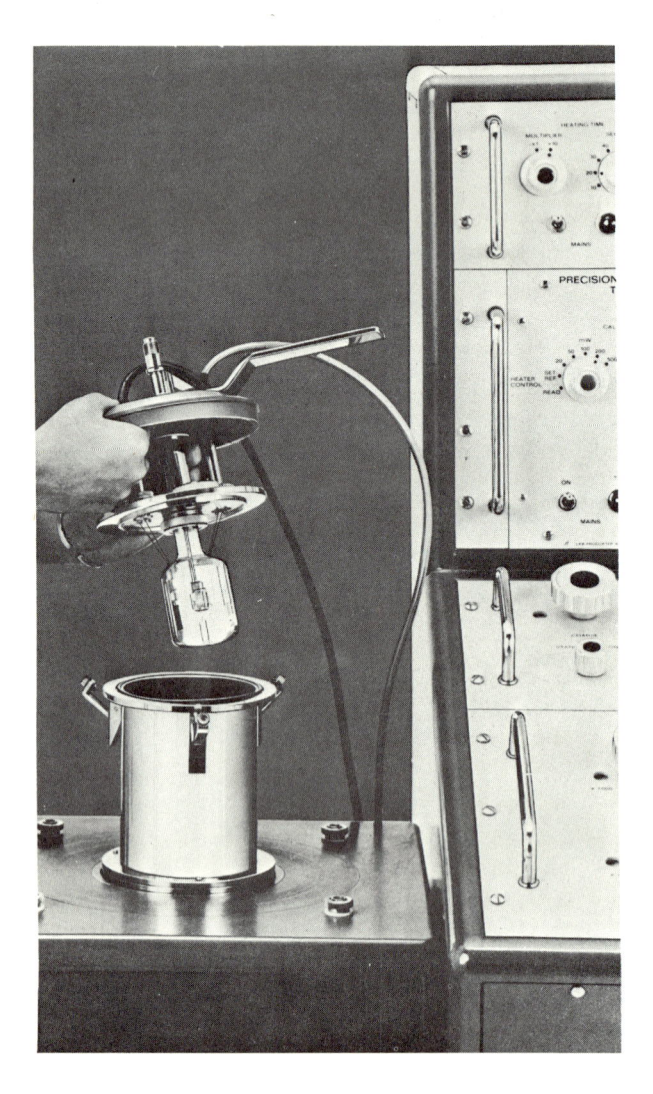

FIG. 25. DETAILED VIEW OF LKB CALORIMETER VESSEL AND CASE. REPRINTED FROM [181] BY COURTESY OF LKB INSTRUMENTS, INC.

1. THERMOMETRIC TITRIMETRY

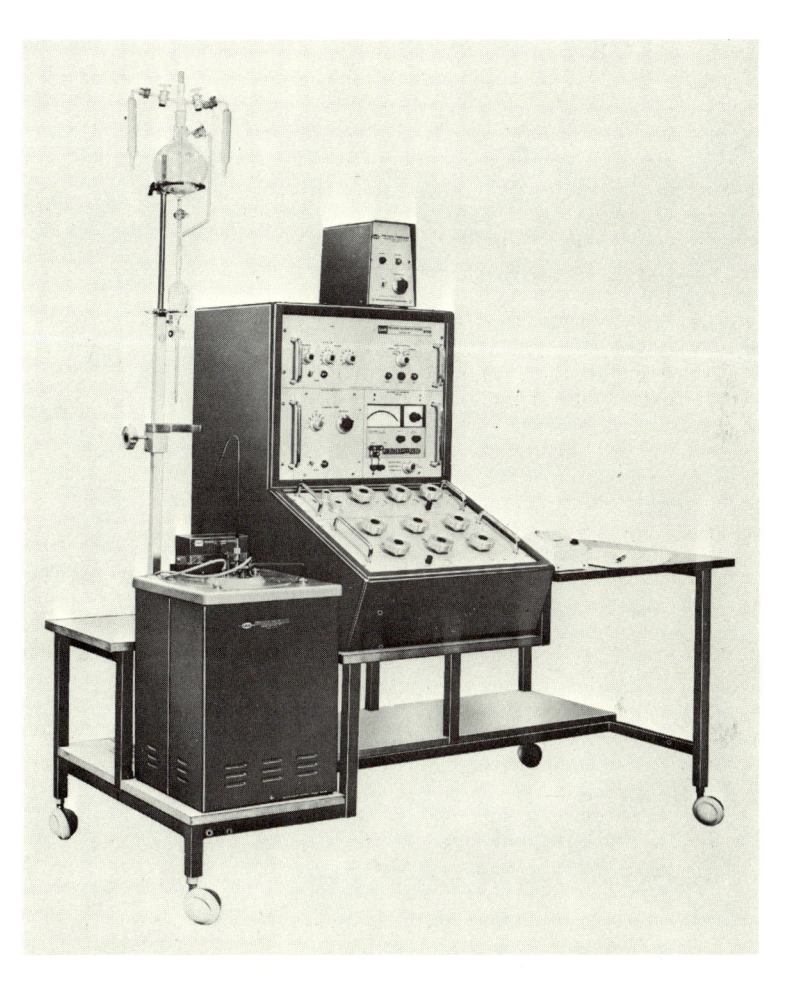

FIG. 26. OVERALL VIEW OF LKB CALORIMETER. REPRINTED FROM [181] BY COURTESY OF LKB INSTRUMENTS, INC.

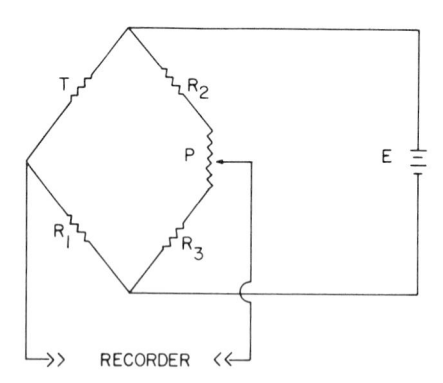

FIG. 27. TEMPERATURE SENSING CIRCUIT. T: THERMISTOR, FENWAL TYPE GB41P2, 10K AT 25°C. R_1, 10K WIREWOUND; R_2 AND R_3, 7.5K WIREWOUND; P, TEN TURN OR MORE POTENTIOMETER. E: TWO 1.4-VOLT MERCURY BATTERIES.

Heathkit or Sargent recorder works well with the above circuit. All other parts shown in Fig. 27 are available from any local electronic parts store. The ten-turn or more potentiometer may be of the ten-turn type with a counting dial (total cost about $16) or a wirewound 22-turn trimming potentiometer (about $1.75). The difference is simply the convenience of the counting dial since the trimming potentiometer must be adjusted with a screwdriver.

Linde et al. [4] have described a buret based on the principle that the rate of flow of a given liquid through a capillary tube is dependent only on pressure and temperature. The buret they constructed (Fig. 28) gave very precise flow rates but required frequent and careful calibration for each different titrant. Lingane [183] has described a constant rate motorized syringe buret. Motorized burets based on this principle are available commercially from Sage Instruments, Inc.

1. THERMOMETRIC TITRIMETRY

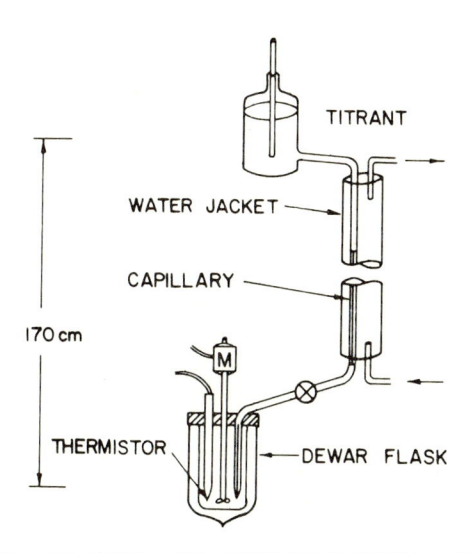

FIG. 28. CONSTANT RATE CAPILLARY BURET. REPRINTED FROM [4, p. 405] BY COURTESY OF THE AMERICAN CHEMICAL SOCIETY.

(about $145), and Greiner Scientific Corporation (about $250). However, a motorized constant rate syringe buret which is equivalent in accuracy to either of the above may be constructed in a few hours and at a cost of approximately $40 [184]. The basic items needed are a Gilmont micrometer syringe and a synchronous motor. A diagram and photograph of the buret are given in Fig. 29 and 30, respectively. Table 14 shows the precision obtainable with this buret. The disadvantages of the buret compared to the commercial models are (a) the difficulty of changing the delivery rate and (b) that only 2.0 and 0.2 ml capacities are available for the Gilmont micrometer syringes.

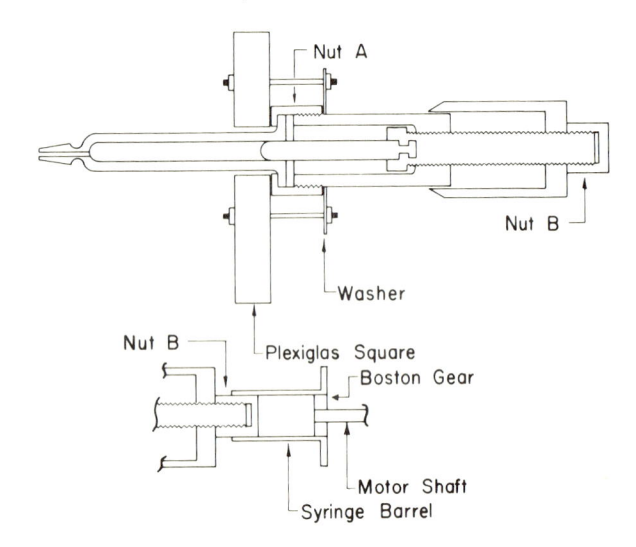

FIG. 29. SCHEMATIC DIAGRAM OF MICROMETER SYRINGE, COLLAR AND DRIVE TRAIN. REPRINTED FROM [184, p. 876] BY COURTESY OF THE DIVISION OF CHEMICAL EDUCATION OF THE AMERICAN CHEMICAL SOCIETY.

FIG. 30. PHOTOGRAPH OF BURET. REPRINTED FROM [184, p. 876] BY COURTESY OF THE DIVISION OF CHEMICAL EDUCATION OF THE AMERICAN CHEMICAL SOCIETY.

1. THERMOMETRIC TITRIMETRY 77

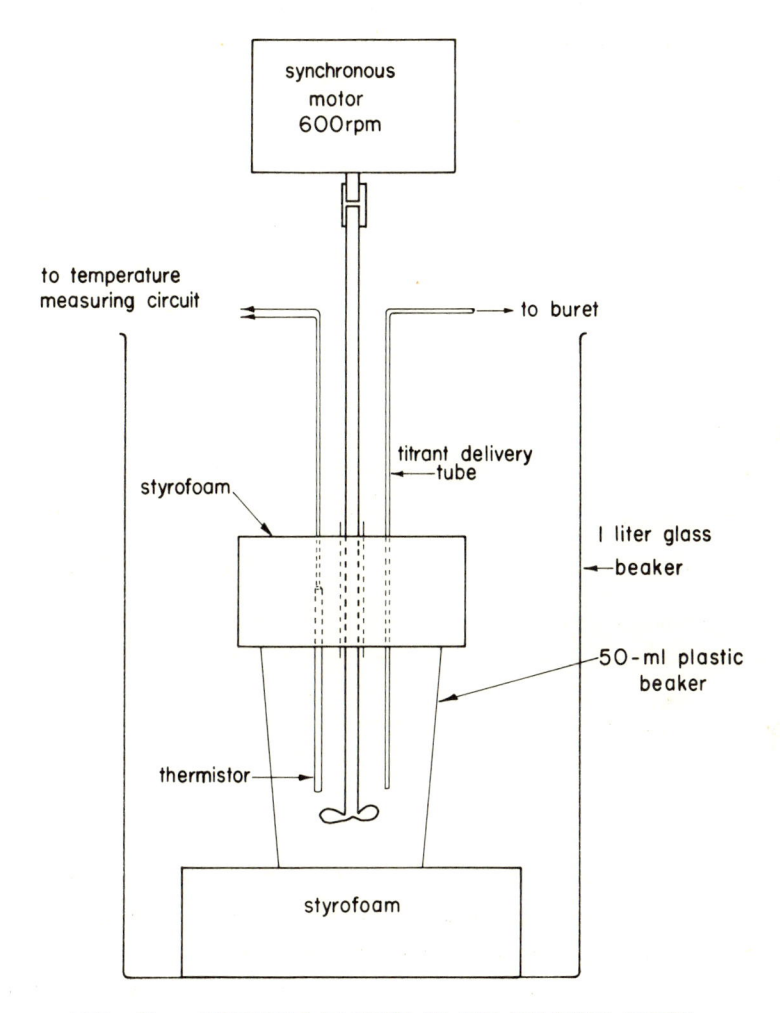

FIG. 31. SCHEMATIC DRAWING OF THE REACTION VESSEL.

TABLE 14[a]

Calibration Data for a Motor Driven Constant Rate Buret

Trial	Delivery rate (mg/sec)
1	35.52
2	35.56
3	35.83
4	35.83
5	35.55
6	35.62
7	35.37
Average	35.61
Average deviation	±0.13 (or 0.4%)
Standard deviation	±0.05

[a]Reprinted from [184, p. 876] by courtesy of the Division of Chemical Education of the American Chemical Society.

A suitable reaction vessel can be made from a 50-ml disposable polypropylene beaker placed on a block of styrofoam inside a 1-liter beaker. The titrant delivery tube and the thermistor probe can be held in place in the beaker by cementing through a small block of styrofoam used for a lid. A schematic drawing of the reaction vessel is shown in Fig. 31.

The approximate cost of the thermometric titration system described above, excluding the recorder, is $75.

REFERENCES

1. J. J. Christensen and R. M. Izatt in *Physical Methods in Advanced Inorganic Chemistry* (H. A. O. Hill and P. Day, Eds), Wiley-(Interscience), New York, 1968, pp. 538-598.
2. J. J. Christensen and R. M. Izatt, Table, "Heats of Proton Ionization and Related Thermodynamic Quantities," in *Handbook of Biochemistry with Selected Data for Molecular Biology* (H. A. Sober, Ed.), Chemical Rubber Publishing Co., Cleveland, Ohio, 1968, pp. J49-J139.
3. H. J. V. Tyrrell and A. E. Beezer, *Thermometric Titrimetry*, Chapman and Hall, Ltd., London, 1968.
4. H. W. Linde, L. B. Rogers, and D. N. Hume, *Anal. Chem.*, 25, 404 (1953).
5. J. L. Ragland, *Soil Sci. Soc. Amer. Proc.*, 26, 133 (1962).
6. G. A. Vaughn and J. J. Swithenbank, *Analyst* (London), 95, 890 (1970).
7. D. N. Hume, *Chem. Eng. News*, Oct. 30, 1961, p. 39.
8. P. G. Zambonin and J. Jordan, *Anal. Chem.*, 41, 437 (1969).
9. J. M. Bell and C. F. Cowell, *J. Amer. Chem. Soc.*, 35, 49 (1913).
10. R. H. Muller, *Ind. Eng. Chem., Anal. Ed.*, 13, 667 (1941).
11. J. Jordan and T. G. Alleman, *Anal. Chem.*, 29, 9 (1957).
12. B. C. Tyson, Jr., W. H. McCurdy, Jr., and C. D. Bricker, *Anal. Chem.*, 33, 1640 (1961).
13. H. J. Keily and D. N. Hume, *Anal. Chem.*, 36, 543 (1964).
14. S. T. Zenchelsky and P. R. Segatto, *Anal. Chem.*, 29, 1856 (1957).
15. P. T. Priestley, *Analyst* (London), 88, 194 (1963).
16. A. B. De Leo and M. J. Stern, *J. Pharm. Sci.*, 54, 911 (1965).
17. G. W. Ewing, *Instrumental Methods of Chemical Analysis*, 2nd ed., McGraw-Hill, New York, 1960, pp. 347-349.
18. J. Jordan, *J. Chem. Educ.*, 40, A5 (1963).

19. H. H. Willard, L. L. Merritt, and J. A. Dean, Instrumental Methods of Analysis, 4th ed., Van Nostrand, Princeton, N.J., 1965, pp. 465-472.
20. J. Jordan, in Treatise on Analytical Chemistry (I. M. Kolthoff and P. J. Elving, Eds.), Part I, Vol. 4, Wiley (Interscience), New York, 1968, pp. 5175-5242.
21. W. W. Wendlandt, Thermal Methods of Analysis, Wiley (Interscience), New York, 1964, pp. 271-296.
22. J. Jordan, in McGraw-Hill Yearbook of Science and Technology (D. I. Eggenberger, Ed.), McGraw-Hill, New York, 1965, pp. 413-415.
23. C. B. Murphy, in Encyclopedia of Industrial Chemical Analysis (F. D. Snell and C. L. Hilton, Eds.), Vol. 3, Wiley (Interscience), New York, 1966, pp. 672-685.
24. P. Papoff and P. G. Zambonin, Ric. Sci., 5, 93 (1965).
25. J. Jordan and R. H. Henry, Microchem. J., 10, 260 (1966).
26. J. P. Phillips, Automatic Titrators, Academic Press, New York, 1959, pp. 110-116.
27. J. Jordan, Rec. Chem. Progr., 19, 193 (1958).
28. J. Jordan, Chimia, 17, 101 (1963).
29. J. Jordan and G. J. Ewing, in Handbook of Analytical Chemistry (L. Meites, Ed.), McGraw-Hill, New York, 1963, Sec. 8, pp. 3-7.
30. S. T. Zenchelsky, Anal. Chem., 32, 289R (1960).
31. L. S. Bark and S. M. Bark, Thermometric Titrimetry, International Series of Monographs in Analytical Chemistry, Vol. 33, Pergamon, New York, 1969.
32. M. Harmelin, Chim. Anal. (Paris), 44, 153 (1962).
33. P. T. Priestley, W. S. Sebborn, and R. F. W. Selman, Analyst. (London), 88, 797 (1963).
34. P. W. Carr, in Critical Review of Analytical Chemistry (Louis Meites, Ed.) Vol. 2, Issue 4, 1972, pp. 491-557.

35. J. J. Christensen, R. M. Izatt, and L. D. Hansen, J. Amer. Chem. Soc., 89, 213 (1967), Ref. 4.
36. J. J. Christensen and R. M. Izatt, J. Phys. Chem., 66, 1030 (1962).
37. L. D. Hansen, unpublished results.
38. B. Sen and W. C. Wu, Anal. Chim. Acta, 46, 37 (1969).
39. J. J. Christensen, R. M. Izatt, L. D. Hansen, and J. D. Hale, Inorg. Chem., 3, 130 (1964).
40. L. D. Hansen, R. M. Izatt, and J. J. Christensen, Inorg. Chem., 2, 1243 (1963).
41. L. G. Sillen and A. E. Martell, Stability Constants of Metal Ion Complexes, 2nd ed., Chem. Soc., Publ. No. 17, 1964.
42. J. J. Christensen, R. M. Izatt, and D. Eatough, Inorg. Chem., 4, 1278 (1965).
43. J. J. Christensen, D. P. Wrathall, and R. M. Izatt, Anal. Chem., 40, 175 (1968).
44. R. M. Izatt, D. Eatough, J. J. Christensen, and C. H. Bartholomew, J. Chem. Soc., A1969, 45.
45. L. D. Hansen and E. A. Lewis, Anal. Chem., 43, 1393 (1971).
46. J. J. Christensen, R. M. Izatt, D. P. Wrathall, and L. D. Hansen, J. Chem. Soc., A1969, 1212
47. R. M. Izatt, J. H. Rytting, L. D. Hansen, and J. J. Christensen, J. Amer. Chem. Soc., 88, 2641 (1966).
48. K. C. Burton and H.M.N.H. Irving, Anal. Chim. Acta, 52, 441 (1970).
49. T. Kiss, Fresenius, Z. Anal. Chem., 252, 12 (1970).
50. H. Weisz, T. Kiss, and D. Klockow, Fresenius, Z. Anal. Chem., 247, 248 (1969).
51. H. Weisz, and T. Kiss, Fresenius, Z. Anal. Chem., 249, 302 (1970).
52. G. A. Vaughan and J. J. Swithenbank, Analyst (London), 90, 594 (1965).

53. M. W. Brown, K. Issa, and A. G. Sinclair, Analyst (London), 94, 234 (1969).
54. J. Jordan and P. W. Carr, in Analytical Calorimetry, (R. S. Porter and J. F. Johnson, Eds.), Plenum, New York, 1968, pp. 203-208.
55. H. J. V. Tyrrell, Talanta, 14, 843 (1967).
56. J. C. Wasilewski, P. T.-S. Pei, and J. Jordan, Anal. Chem., 36, 2131 (1964).
57. C. A. Reynolds and M. J. Harris, Anal. Chem., 41, 348 (1969).
58. P. T. Priestley, W. S. Sebborn, and R. F. W. Selman, Analyst (London), 90, 589 (1965).
59. P. T. Priestley, J. Sci. Instr., 42, 35 (1965).
60. D. P. Wrathall, R. M. Izatt, and J. J. Christensen, J. Amer. Chem. Soc., 86, 4779 (1964).
61. J. Jordan and W. H. Dumbaugh, Jr., Anal. Chem., 31, 210 (1959).
62. A. B. De Leo and M. J. Stern, J. Pharm. Sci., 53, 993 (1964).
63. J. Jordan, P. T. Pei, and R. A. Javick, Anal. Chem., 35, 1534 (1963).
64. N. D. Weiner and A. Felmeister, Anal. Chem., 38, 515 (1966).
65. F. E. Freeberg, Anal. Chem., 41, 54 (1969).
66. J. Jordan and M. P. Ben-Yair, Ark. Kemi, 11, 239 (1957).
67. J. P. Gallet and R. A. Paris, Anal. Chim. Acta, 39, 181 (1967).
68. J. P. Gallet and R. A. Paris, Anal. Chim. Acta, 40, 321 (1968).
69. J. Brandstetr, Fresenius' Z. Anal. Chem., 254, 34 (1971).
70. F. J. Miller and P. F. Thomason, Talanta, 2, 109 (1959).
71. F. J. Miller and P. F. Thomason, Anal. Chem., 31, 1498 (1959).

72. R. A. Paris and J. Vial, Chim. Anal. (Paris), 34, 3 (1952).
73. R. D. Daftary and B. C. Haldar, Anal. Chim. Acta, 25, 538 (1961).
74. R. E. Press, Tappi, 48, 464 (1965).
75. N. D. Jespersen and J. Jordan, Anal. Letters, 3, 323 (1970).
76. J. Jordan and E. J. Billingham, Jr., Anal. Chem., 33, 120 (1961).
77. J. L. Rasmussen and T. Nielsen, Acta Chem. Scand., 17, 1623 (1963).
78. J. J. Christensen, R. M. Izatt, J. D. Hale, R. T. Pack, and G. D. Watt, Inorg. Chem., 2, 337 (1963).
79. W. L. Everson and E. M. Ramirez, Anal. Chem., 39, 1771 (1967).
80. W. L. Everson, Anal. Chem., 43, 201 (1971).
81. P. W. Carr, Anal. Chem., 43, 756 (1971).
82. R. M. Izatt, H. D. Johnston, and J. J. Christensen, Brigham Young Univ., unpublished results.
83. F. J. Miller and P. F. Thomason, Anal. Chim. Acta, 21, 112 (1959).
84. E. J. Billingham, Jr., and A. H. Reed, Anal. Chem., 36, 1148 (1964).
85. W. A. Alexander, C. J. Mash, and A. McAuley, Analyst (London), 95, 657 (1970).
86. E. Schafer and E. Wilde, Fresenius' Z. Anal. Chem., 130, 396 (1950).
87. T. Takeuchi and M. Yamazaki, Chem. Abstr., 72, 42997t (1970).
88. W. L. Everson, Anal. Chem., 36, 854 (1964).
89. W. L. Everson and E. M. Ramirez, Anal. Chem., 37, 812 (1965).

90. W. L. Everson and E. M. Ramirez, Anal. Chem., 37, 806 (1965).
91. T. R. Crompton and B. Cope, Anal. Chem., 40, 274 (1968).
92. R. D. Parker and T. Vlismas, Analyst (London), 93, 330 (1968).
93. J. Belisle, Anal. Chim. Acta, 54, 156 (1971).
94. S. T. Zenchelsky, J. Periale, and J. C. Cobb, Anal. Chem., 28, 67 (1956).
95. C. J. Quilty, Anal. Chem., 39, 666 (1967).
96. K. L. Snyder, Chem. Eng. Progr., 64, 75 (1968).
97. T. E. Mead, J. Phys. Chem., 66, 2149 (1962).
98. E. J. Forman and D. N. Hume, Talanta, 11, 129 (1964).
99. L. H. Greathouse, H. J. Janssen, and C. H. Haydel, Anal. Chem., 28, 357 (1956).
100. K. V. Topchieva, I. F. Moskovskaya, and N. A. Dobrokhotova, Chem. Abstr., 62, 59g (1965).
101. J. Valcha, Chem. Abstr., 63, 17191d (1966).
102. C. Jambon and J. C. Merlin, C. R. Acad. Sci., Paris, Ser. C, 272, 195 (1971).
103. V. J. Vajgand, F. F. Gaal, and S. S. Brusin, Talanta, 17, 415 (1970).
104. J. Jordan, J. Meier, E. J. Billingham, Jr., and J. Pendergrast, Nature, 187, 318 (1960).
105. J. Jordan, J. Meier, E. J. Billingham, Jr., and J. Pendergrast, Anal. Chem., 31, 1439 (1959).
106. J. Jordan, J. Meier, and E. J. Billingham, Jr., Anal. Chem., 32, 651 (1960).
107. J. H. Forsberg and T. Moeller, Inorg. Chem., 8, 889 (1969).
108. R. C. Paul, R. Parkash, S. C. Ahluwalia, and S. S. Sandhu, Indian J. Chem., 6, 373 (1968).
109. S. T. Zenchelsky and P. R. Segatto, J. Amer. Chem. Soc., 80, 4796 (1958).

1. THERMOMETRIC TITRIMETRY

110. A. Kettrup and H. Specker, Fresenius' Z. Anal. Chem., 230, 241 (1967).
111. I. Grenthe and I. Leden, in Proc. 8th Int. Conf. Coord. Chem. (V. Gutmann, Ed.), Springer-Verlag, New York, 1964, p. 332.
112. R. Arnek, Acta Chem. Scand., 23, 1986 and references therein (1969).
113. C. D. Schmulbach and D. M. Hart, J. Amer. Chem. Soc., 86, 2347 (1964).
114. W. A. Alexander, C. J. Mash, and A. McAuley, Talanta, 16, 535 (1969).
115. A. Brenner, J. Electrochem. Soc., 112, 611 (1965).
116. W. H. Metzger, Jr., A. Brenner, and H. I. Salmon, J. Electrochem. Soc., 114, 131 (1967).
117. A. Brenner and W. H. Metzger, Jr., J. Electrochem. Soc., 115, 258 (1968).
118. J. J. Christensen, J. Ruckman, D. J. Eatough, and R. M. Izatt, Thermochim. Acta, 3, 203 (1972).
119. D. J. Eatough, J. J. Christensen, and R. M. Izatt, Thermochim. Acta, 3, 219 (1972).
120. D. J. Eatough, R. M. Izatt, and J. J. Christensen, Thermochim. Acta, 3, 233 (1972).
121. J. J. Christensen, R. M. Izatt, L. D. Hansen, and J. A. Partridge, J. Phys. Chem., 70, 2003 (1966).
122. A. C. M. Wanders and T. N. Zwietering, J. Phys. Chem., 73, 2076 (1969).
123. R. M. Izatt, D. Eatough, R. L. Snow, and J. J. Christensen, J. Phys. Chem., 72, 1208 (1968).
124. L. G. Sillen, Acta Chem. Scand., 16, 159 (1962).
125. P. Paoletti, A. Vacca, and D. Arenare, J. Phys. Chem., 70, 193 (1966).
126. W. C. Davidon, Argonne National Lab. Rep. ANL 5990 Rev 2, Argonne, Ill., 1966.

127. R. M. Izatt, L. D. Hansen, J. H. Rytting, and J. J. Christensen, J. Amer. Chem. Soc., 87, 2760 (1965).
128. J. J. Christensen, D. P. Wrathall, Reed M. Izatt, and D. O. Tolman, J. Phys. Chem., 71, 3001 (1967).
129. H. H. Stroh and G. Westphal, Chem. Ber., 96, 184 (1963).
130. A. C. Cumming, Z. Phys. Chem. Abt. A, 57, 574 (1907).
131. P. Leggate and G. E. Dunn, Can. J. Chem., 43, 1158 (1965).
132. P. O. Lumme, Suomen Kemistilehti, B, 30, 176 (1967).
133. A. M. Liquori and A. Ripamonti, Gazz, Chim. Ital., 85, 578 (1955).
134. S. Kilpi and P. Harjanne, Suomen Kemistilehti, B, 21, 14 (1948).
135. K. Winkelblech, Z. Phys. Chem. Abt. A, 36, 564 (1901).
136. D. Peltier and M. Conti, Compt. Rend. Ser. C, 244, 2811 (1957).
137. C. G. Clear and G. E. K. Branch, J. Org. Chem., 2, 522 (1938).
138. A. Bryson and R. W. Matthews, Australian J. Chem., 10, 128 (1957).
139. R. A. Robinson and A. I. Biggs, Australian J. Chem., 10, 128 (1957).
140. J. Johnston, Z. Phys. Chem. Abt. A, 57, 557 (1907).
141. A. Albert and R. Goldacre, Nature, 149, 245 (1942).
142. L. D. Hansen, B. D. West, E. J. Baca, and C. L. Blank, J. Amer. Chem. Soc., 90, 6588 (1968).
143. P. Papoff, G. Torsi, and P. G. Zambonin, Gazz. Chim. Ital., 95, 1031 (1965).
144. C. E. Vanderzee and A. S. Quist, J. Phys. Chem., 65, 118 (1961).
145. R. D. McCoy and D. F. Swinehart, J. Amer. Chem. Soc., 76, 4708 (1954).

146. R. K. Murman and F. Basolo, J. Amer. Chem. Soc., 77, 3484 (1955).
147. H. C. Brown, D. H. McDaniel, and O. Haflinger in Determination of Organic Structures by Physical Methods (E. A. Braude and F. C. Nachod, eds.), Vol. 1, Academic Press, New York, 1955, p. 581.
148. A. H. M. Kirby and A. Neuberger, Biochem. J., 32, 1146 (1938).
149. S. P. Datta and A. K. Grzybowski, J. Chem. Soc., B 1966, 136.
150. R. G. Bates and G. D. Pinching, J. Res. Nat. Bur. Stand., 43, 519 (1949).
151. R. G. Bates and H. B. Hetzer, J. Phys. Chem., 65, 667 (1961).
152. S. P. Datta, A. K. Grzybowski, and B. A. Weston, J. Chem. Soc., (London), 1963, 792.
153. S. P. Datta and A. K. Grzybowski, Trans. Faraday Soc., 54, 1179 (1958).
154. E. J. King, J. Amer. Chem. Soc., 73, 155 (1951).
155. B. B. Owen, J. Amer. Chem. Soc., 56, 24 (1934).
156. J. J. Christensen, D. P. Wrathall, J. O. Oscarson, and R. M. Izatt, Anal. Chem., 40, 1713 (1968).
157. R. J. Raffa, M. J. Stern, and L. Malspeis, Anal. Chem., 40, 70 (1968).
158. S. J. Gill and E. L. Farquhar, J. Amer. Chem. Soc., 90, 3039 (1968).
159. T. D. Epley and R. S. Drago, J. Amer. Chem. Soc., 89, 5770 (1967).
160. D. Neerinck, A. Van Audenhaege, L. Lamberts, and P. Huyskens, Nature, 218, 461 (1968).
161. G. Olofsson, Acta Chem. Scand., 21, 1887, 1892, 2143 (1967).

162. T. H. Benzinger, Proc. Nat. Acad. Sci. U. S., 42, 109 (1956).
163. P. Paoletti, A. Vacca, and A. Dei, in Progress in Coordination Chemistry (M. Cais, Ed.), Elsevier, New York, 1968, p. 631.
164. F. Becker and H. M. Luschow, in Proc., 8th Int. Conf. Coord. Chem. (V. Gutmann, Ed.), Springer-Verlag, New York, 1964, p. 334.
165. R. M. Izatt, D. Eatough, J. J. Christensen, and C. H. Bartholomew, J. Chem. Soc., A1969, 47.
166. F. Becker and R. Grundmann, Z. Phys. Chem. (Frankfurt), 66, 137 (1969).
167. P. C. Harris and T. E. Moore, Inorg. Chem., 7, 656 (1968).
168. R. M. Izatt, D. P. Nelson, J. H. Rytting, B. L. Haymore, and J. J. Christensen, J. Amer. Chem. Soc., 93, 1619 (1971).
169. G. J. Papenmeier and J. M. Campagnoli, J. Amer. Chem. Soc., 91, 6579 (1969).
170. D. P. Wrathall and W. L. Gardner in Temperature: Its Measurement and Control in Science and Industry, (H. H. Plumb, Ed.), Vol. 4, Part 3, 1973, p. 2223.
171. D. J. Eatough, Anal. Chem., 42, 635 (1970).
172. T. F. Bolles and R. S. Drago, J. Amer. Chem. Soc., 87, 5015 (1965).
173. F. Becker, J. Barthel, N. G. Schmahl, G. Lange, and H. M. Luschow, Z. Phys. Chem. (Frankfurt am Main), 37, 33 (1963).
174. F. Becker, J. Barthel, N. G. Schmahl and H. M. Luschow, Z. Phys. Chem. (Frankfurt am Main), 37, 52 (1963).
175. W. C. Duer and G. L. Bertrand, J. Amer. Chem. Soc., 92, 2587 (1970).

176. P. Papoff, G. Torsi, and P. G. Zambonin, Gazz. Chim. Ital., 95, 1031 (1965).
177. S. Cabani and P. Gianni, J. Chem. Soc., A1968, 547.
178. J. J. Christensen, J. H. Rytting, and R. M. Izatt, J. Chem. Soc., A1969, 861.
179. J. J. Christensen, R. M. Izatt, and L. D. Hansen, Rev. Sci. Instr., 36, 779 (1965).
180. J. J. Christensen, H. D. Johnston, and R. M. Izatt, Rev. Sci. Instr., 39, 1356 (1968).
181. LKB Instrument Company, "The Precision Calorimetry System LKB 8700," LKB-Produkter AB, S-161 25 Bromma 1, Sweden.
182. L. D. Hansen, D. Kenney, W. M. Litchman, and E. A. Lewis, J. Chem. Educ., 48, 851 (1971).
183. J. J. Lingane, Anal. Chem., 20, 285 (1948).
184. L. D. Hansen, W. M. Litchman, E. A. Lewis, and R. E. Allred, J. Chem. Educ., 46, 876 (1969).

CHAPTER 2

SOME UNUSUAL END-POINT DETECTION
METHODS INVOLVING HETEROGENEOUS PROCESSES

D. J. Curran

Department of Chemistry
University of Massachusetts
Amherst, Massachusetts

I.	INTRODUCTION	92
II.	PRESSUREMETRIC TITRATIONS	93
	A. Introduction	93
	B. Theory	94
	C. Apparatus	97
	D. Applications	113
III.	CRYOSCOPIC TITRATIONS	123
	A. Introduction	123
	B. Theory and Methods	123
	C. Apparatus and Experimental Methods	128
	D. Applications	130
IV.	PHASE TITRATIONS	135
	A. Introduction	135
	B. Discussion of Methods	136
	C. Apparatus and Experimental Methods	153
	D. Applications	155
V.	FLAME PHOTOMETRIC TITRATIONS	172
	A. Introduction	172
	B. Alkaline Earth-Phosphorus Systems	174
	C. The Lanthanum-Phosphate System	176

D. Conclusions 177
REFERENCES 178

I. INTRODUCTION

This chapter presents four different methods of titration end-point detection: pressuremetric titrations, cryoscopic titrations, phase titrations, and flame photometric titrations. Together they cover a considerable variety of chemistry. Thermistors, the human eye, and pressure transducers are used for end-point detection. Acid-base, redox, complexation, and precipitation reactions are all encountered; aqueous and nonaqueous solvents appear. In short, these four methods manage to cover the broad scope of present day activity in titrimetric methods. However, the underlying principles, operating procedures, and interpretation of data of these methods are each quite diverse and there is no doubt that each deserves classification as a separate method of titration, as distinct as, for example, the potentiometric, amperometric, and spectrophotometric methods.

A search for a common theme among these methods reveals that they all require the presence in one way or another of more than one phase in the system. For pressuremetric titrations these phases are gas and solution; for cryoscopic titrations, solid and solution; for phase titrations, two different solutions. For flame photometric titrations the species giving rise to the measured emission intensity must undergo the following phase transfers: solution to solid to gaseous. It is safe to say that the inclusion of heterogeneous processes is a discernible trend in current research on titrimetric methods.

2. UNUSUAL END-POINT DETECTION METHODS

II. PRESSUREMETRIC TITRATIONS

A. Introduction

When a titration reaction which involves the production or consumption of gas is carried out in a closed vessel containing both solution and gas phases, the system pressure can be monitored to detect the end point. A number of older methods of analysis, usually termed gasometric, are based on a determination of the total amount of gas involved in the chemical reaction. In a system containing the two phases, such a measurement scheme involves problems of gas solubility and gas separation. The remedy has been to work at reduced pressures. With the titration technique, a quantitative measurement of the total amount of gas generated or consumed is not needed and the requirement for vacuum systems is eliminated. The titration can be carried out at gauge pressures and in the presence of other gases. It is useful also to compare pressuremetric titrations with volumetric titrations in terms of the amount of material required. A typical volumetric titration uses in the vicinity of 5 mmoles of titrate. For a gas phase volume of 50 ml held at 300.0°K, the addition or removal of this amount of an ideal gas will result in a pressure change of 2.5 atm. Not only is the measurement of this pressure change within the capabilities of modern pressure transducer systems, but a pressure change two orders of magnitude smaller is also easily measured. Thus, pressuremetric titrations can be performed routinely at the milligram level of sought-for-constituent. The apparatus is no more complex to operate than that required for a potentiometric titration and the data are in the same form: transducer output in volts versus volume of titrant added, or time in the case of coulometric generation of titrant. Unlike potentiometric titrations, the titration curves are of the linear-segmented type and resemble those of amperometric titrations with one polarized electrode, and those of some spectrophotometric titrations.

B. Theory

The theory of pressuremetric titration curves has been considered by Curran and Driscoll [1] and Driscoll [2] for volumetric addition of titrant, and by Kronfeld [3] and Curran and Curley [4] for coulometric addition of titrant. The approach of these authors is followed here. For simplicity, the coulometric case is examined and the results extended to the volumetric case.

Consider a closed vessel held at constant temperature and equipped with connections to a pressure measuring device, magnetic stirring, and a pair of electrodes. Titrant is generated coulometrically at one of the electrodes and the counter electrode is either electrochemically isolated or, if dipping directly into the titrant solution, the reaction at it has no influence on the titration reaction or the end-point detection. The titration reaction may be represented by the following:

$$aA + bB \rightleftharpoons gG(aq) + pP \quad (1)$$

where A is the unknown, B is the titrant, and G is a gaseous product. The moles of gas generated by the titration reaction, n_{gg}, is related to the moles of titrant consumed, n_t, by the stoichiometric ratio

$$n_{gg}/n_t = g/b = \alpha \quad (2)$$

The moles of titrant are given by Faraday's law, where i is the current in amperes, t is the electrolysis time in seconds, N is the number of electrons involved in the electrode process, and F is the value of the faraday. Therefore,

$$n_{gg} = \alpha it/NF \quad (3)$$

2. UNUSUAL END-POINT DETECTION METHODS

It remains to show that the pressure signal is proportional to the time of electrolysis. It is assumed that the gas and solution phases are ideal and that the titration reaction goes quantitatively to the right. Concentrations and partial pressures will be used in place of activities and fugacities, respectively, and it is assumed that the volumes of the gas phase, V_g, and the solution phase, V_a, are constant. It is also assumed that the gas produced by the titration reaction partitions between the phases according to

$$G(aq) \rightleftharpoons G(gas) \quad (4)$$

It is convenient to consider the partition process in terms of Henry's law and write

$$K = p/Y \quad (5)$$

where K is the Henry law constant, p is the partial pressure of the gas, and Y is the mole fraction of the gas in the aqueous phase. Upon generation of gas, the ratio of the change in partial pressure, Δp, to the change in mole fraction, ΔY, is given by the same constant:

$$K = \Delta p/\Delta Y \quad (6)$$

To a very good approximation, the sum, k, of the moles of all species present in solution does not change appreciably during the titration and the mole fraction of dissolved gas may be written as

$$Y = n_a/k \text{ or } \Delta Y = \Delta n_a/k \quad (7)$$

where n_a is the moles of gas in the aqueous phase. Therefore the expression for K may be given as

$$K = \Delta p k/\Delta n_a \quad (8)$$

Now the total moles of gas produced by the titration reaction, n_{gg}, is the sum of that found in the aqueous phase, n_a, and that in the gas phase, n_g, or

$$n_{gg} = n_a + n_g \text{ and } \Delta n_{gg} = \Delta n_a + \Delta n_g \quad (9)$$

At constant volume and temperature, the change in partial pressure of the gas is given by the ideal gas law:

$$\Delta p = \Delta n_g RT/V_g \tag{10}$$

Solving Eq. (8) and (10) for Δn_a and Δn_g, respectively, and substituting in Eq. (9)

$$n_{gg} = \Delta p k/K + \Delta p V_g/RT \tag{11}$$

$$n_{gg} = \Delta p(k/K + V_g/RT) \tag{12}$$

Substituting Eq. (3) into Eq. (12) and rearranging

$$\Delta p = \frac{\alpha i t}{NF} \frac{KRT}{(kRT + KV_g)} \tag{13}$$

For a given experiment, all of the quantities on the right-hand side of Eq. (13) are constant except t, and

$$\Delta p = K_1 t \tag{14}$$

Equation (14) is valid up to and including the equivalence point. Beyond the equivalence point, no more gas is produced and

$$\Delta p = K_1 t_{ep} = K_2 \tag{15}$$

where t_{ep} is the time required to reach the equivalence point (ep). In practice, t_{ep} is obtained from the intersection of the lines drawn through the data points obtained before and after the equivalence point. The moles of titrant required to reach the equivalence point is then calculated from Faraday's law and the moles of unknown obtained from this result and the stoichiometric ratio of moles of unknown to moles of titrant given by Eq. (1):

$$\text{moles of unknown} = (a/b)(\text{moles of titrant}) \tag{16}$$

For pressure transducer systems with electrical readout directly proportional to pressure, Eqs. (14) and (15) become

$$E = K_3 t \tag{17}$$

$$E = K_4 t_{ep} = K_5 \tag{18}$$

Equation (17) predicts a zero intercept for a plot of E vs t. This will only be true when the initial pressure in the reactor is identical to the reference pressure for the transducer system in use. Otherwise, a constant is introduced in Eq. (17). Further, the transducer output may be a linear function of pressure but this functional relationship could also have a nonzero intercept and a constant could appear in Eq. (17) from this source. It should also be mentioned that if the partial pressure of the gas to be generated is initially zero, Henry's law is not immediately applicable and Eq. (17) cannot be expected to hold early in the titration. From a theoretical point of view, there is an advantage in keeping the total pressure change small during the course of the titration. With this condition, the equation of state for the gas can no doubt be represented by a linear relationship over the limited range of pressure change even for nonideal gases.

To extend the arguments to volumetric addition of titrant, it is noted that

$$n_{gg} = \alpha C_t V_t \qquad (19)$$

where C_t is the concentration of titrant in moles per liter, V_t is the volume of titrant added in liters, and α has been defined. If the gas phase volume remains constant and dilution is negligible, then it follows that

$$\Delta p = K'_1 V_t \qquad (20)$$

The above conditions are easily obtained as will be shown in the section on apparatus. Curran and Driscoll [1] have shown that linear-segmented titration curves are still predicted when the gas phase volume changes due to titrant addition, but the disadvantage is that the pressure signal due to compression of the gas phase can be very much larger than the pressure signal due to gas generation.

C. Apparatus

A schematic diagram of the apparatus required for pressuremetric titrations is shown in Fig. 1. The dotted line encloses those units that may have to be thermostated. Some method of titrant delivery is necessary and it may be either

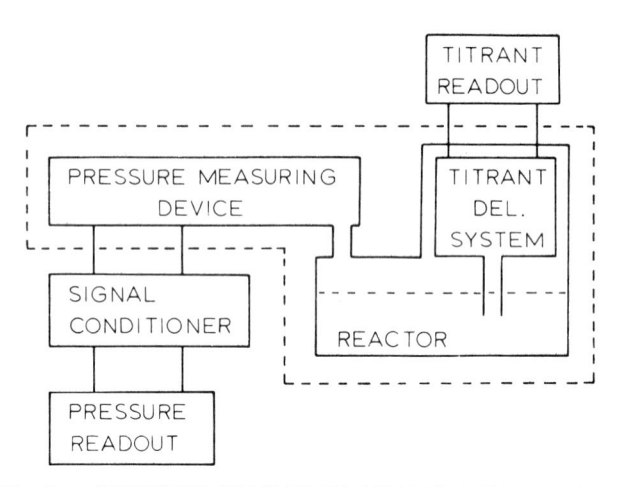

FIG. 1. SCHEMATIC DIAGRAM OF APPARATUS FOR PRESSURE-METRIC TITRATIONS.

volumetric or coulometric. In the former case, the buret tip may or may not be immersed in the solution. It would appear that almost any method of pressure measurement could be used. The signal conditioner would be required only if the pressure signal is transduced to an electrical one. A detailed discussion of modern pressure transducers is beyond the scope of this presentation since practically every conceivable transducing principle has been used. The reader is referred to books by Lion [5] and Neubert [6] and a review article by Curran [7] for information.

Perhaps the simplest and certainly the most obvious choice of pressure measuring device is the manometer. An apparatus using this idea and volumetric addition of titrant was described by Gottlieb [8], and is shown in Fig. 2. The reactor is a 50-ml round-bottom flask equipped with a bar magnet for stirring. The tip of a 10-ml buret and a capillary

2. UNUSUAL END-POINT DETECTION METHODS   99

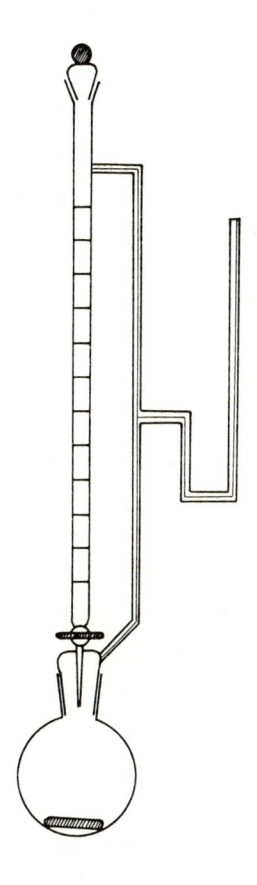

FIG. 2. APPARATUS FOR PRESSUREMETRIC TITRATIONS ACCORDING TO GOTTLIEB [8]. BY PERMISSION.

tube are glassblown to the hollow glass stopper of the flask as shown. The upper end of the capillary tubing joins the upper end of the buret at a point above the solution level.

The buret is sealed against atmospheric pressure by a second hollow glass stopper. A small hole is bored through the glass wall of the buret and another through the stopper so the system can be opened or closed by a rotatory motion of the stopper. Finally, a capillary U-tube containing a colored liquid is attached to the capillary side arm. A bore of 1 mm was said to be suitable for titrations involving 0.05 to 0.5 N solutions. The virtue of this apparatus is its simplicity. In operation the system is usually returned to atmospheric pressure after addition of each drop of titrant as the end point is approached. There will be a relative change in behavior of the liquid in the U-tube manometer as one goes through the end point and the operation is directly analogous to the use of a colored indicator. Gottlieb later modified the apparatus to obtain data suitable for plotting titration curves [9]. The arm of the U-tube not adjacent to the capillary side arm was shortened and a buret inverted over it. The inverted buret is filled with water and the column of water is supported by a beaker of water fitted over the lower half of the modified U-tube. This arrangement removed the tediousness of the earlier drop-by-drop approach to the end point. A 20-ml inverted buret was suitable for titrations involving 0.1 N solutions. The volumetric displacement of the water in the buret was measured as a function of the milliliters of titrant added. The apparatus is closed to the atmosphere over the entire duration of the titration. With efficient stirring, it was found that 1 to 2 min. were sufficient to reach a stable pressure reading after the addition of 1 ml of 0.1 N titrant. Using three or four data points before the end point and two to three after, the total time required for a titration was 6 to 12 min.

In a still later modification, Gottlieb replaced the U-tube manometer with a horizontal column of water [10]. The

2. UNUSUAL END-POINT DETECTION METHODS

titrant buret was sealed with a solid glass stopper and a three-way stopcock was blown on near the top of the buret and opposite the capillary side arm shown in Fig. 2. One of the two remaining outlets of the stopcock went to a suction line and the other was joined to a horizontal length of glass tubing which was bent at a right angle near the free end. This end of the tubing dipped into a beaker of water. The horizontal section was backed by a movable millimeter or 0.5 mm scale. With the stopcock open to the buret, horizontal tubing, and the suction line, water is drawn into the horizontal tubing. After turning the stopcock to leave only the horizontal tubing and the buret connected, the scale is adjusted so the extremity of the water column is at zero on the scale. Gas evolution or consumption in the reactor results in a displacement of the water column which is read in millimeters. A titration curve is constructed by plotting displacement versus volume of titrant delivered. Diameters of 5, 2, and 1 mm for the horizontal tubing were recommended for titrating 0.1, 0.01, and 0.001 N solutions, respectively. There are two main drawbacks to the equipment described by Gottlieb: the sensitivity of the pressure measurement is limited to the ability of the eye to see a change in the position of the liquid in the capillary, and the glassware is somewhat awkward to thermostat. Both of these points become significant when titrating dilute solutions.

An apparatus which overcomes these difficulties has been reported by Curran and Swarin [11]. The liquid level in one arm of a manometer is detected by a pair of conductivity electrodes and operational amplifier circuitry is used to produce a readout which is linear with liquid level and therefore linear with pressure. Figure 3 shows the manometer arrangement. A solution, 0.03 wt% in Triton X-100 and 5×10^{-4} M in KCl, was used as the manometer fluid. Port P_2 is connected to

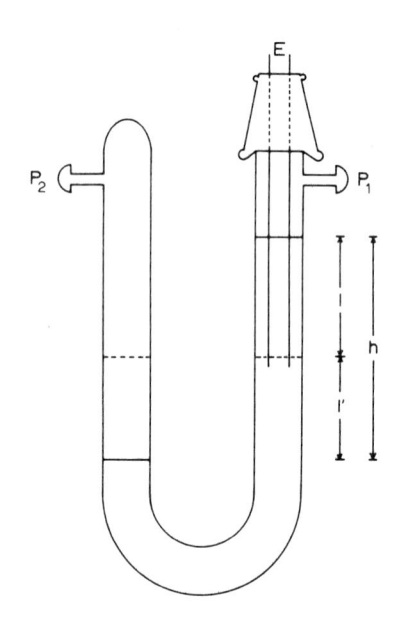

FIG. 3. ELECTROLYTE-FILLED MANOMETER. ----, INITIAL LEVEL OF MANOMETER FLUID. ———, FINAL LEVEL OF MANOMETER FLUID. E, CONDUCTIVITY ELECTRODES. REPRODUCED FROM REF. [11] WITH PERMISSION.

the reactor while port P_1 is either open to the atmosphere or connected to a closed reference vessel. Platinum foil electrodes, 5 x 100 x 0.76 mm were epoxied to 9 x 190 x 2 mm glass plates and platinum wire leads were spot welded to one end of the foils. The electrodes were mounted facing each other using 5 mm lengths of glass tubing as spacers and the leads were brought through a ℥ 29/42 female glass joint and epoxied in place. In operation, the manometer fluid is not leveled; rather the height difference between the arms is allowed to

2. UNUSUAL END-POINT DETECTION METHODS

vary continuously. Goldstein [12] has called this a free manometer technique and a consequence is that both volume and pressure vary in the system. Some sensitivity in pressure measurement is necessarily lost but Curran and Swarin [11] have shown that even when manometer tubing with an internal diameter of 1.5 cm is used for the device, the height change between arms is a linear function of pressure. Further, a change in liquid level of 30 μin. could be detected so the loss in sensitivity is not serious in practical terms for titration work. Figure 3 also shows the total height difference, h, between arms as the sum of l', the height change on the left-hand side, and l, the height change of the right-hand side. Due to the volume occupied by the portion of the electrode assembly immersed in the manometer fluid, l is not equal to l'. This effect does not destroy the linearity of response provided the horizontal cross-sectional area of the electrode assembly is vertically uniform.

The circuit diagram for the system and its circuit components are shown in Fig. 4. Amplifier 2 is wired as an amplitude stabilized sine wave generator. A 500 Hz, 18 V peak-to-peak signal appears across potentiometer P_2 where it is attenuated to about 1 V before entering amplifier 1, which is a follower amplifier used to isolate the cell from the sine wave generator. The output of the follower enters the cell which acts as the input impedance for amplifier 3 which is wired as a simple inverter amplifier. The outputs of the follower amplifier and amplifier 3 are summed by amplifier 4 which is also wired in the inverter configuration. It should be noted that these two signals are 180 deg out of phase and that the summation is therefore a subtraction. The output of amplifier 4 is a signal whose amplitude is proportional to the difference in conductance between that of the cell and that of potentiometer P_3. P_3 is used to null out any initial cell

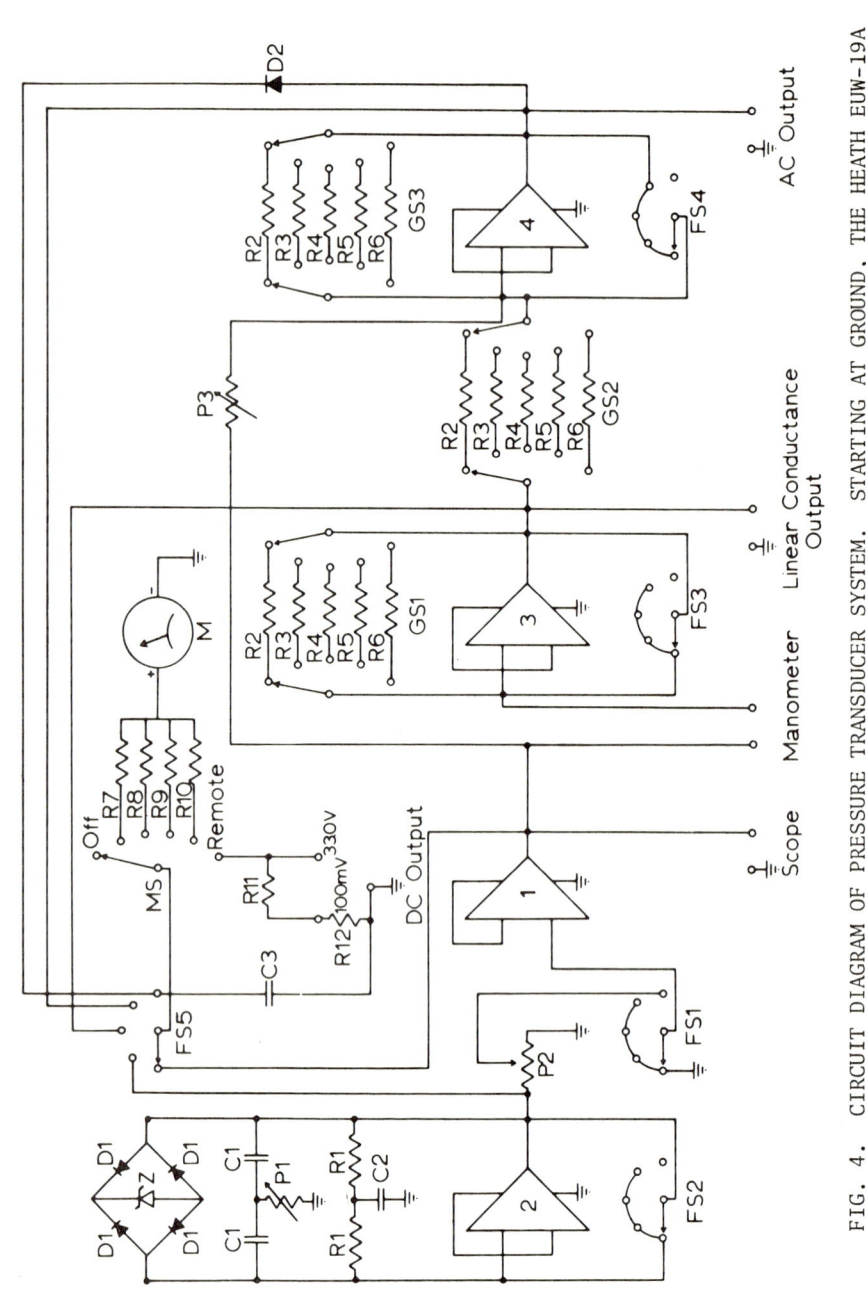

FIG. 4. CIRCUIT DIAGRAM OF PRESSURE TRANSDUCER SYSTEM. STARTING AT GROUND, THE HEATH EUW-19A AMPLIFIERS HAVE PINS NUMBERED CLOCKWISE AS 4, 5, 1, 2, AND 3. THEY ARE RESPECTIVELY, GROUND, NON-INVERTING INPUT, INVERTING INPUT, BALANCE SUPPLY, AND OUTPUT. FROM REF. [11] BY PERMISSION.

List of Components in Fig. 4

D1	In 4002 Texas Instruments	R4	10 kΩ
D2	In 538 Texas Instruments	R5	100 kΩ
Z	9.0 V, 1 W Zener	R6	1 MΩ
C1	0.0022 µF, 10%, paper	R7	20 kΩ
C2	0.0047 µF, 10%, paper	R8	100 kΩ
C3	0.22 µF, 10%, paper	R9	200 kΩ
R1	154 kΩ, all are 1/2 W,	R10	300 kΩ
R2	100 Ω, carbon film	R11	992 kΩ
R3	1 kΩ	R12	3.9 kΩ
P1	100 kΩ, 0.5% linear, 10-turn wire wound potentiometer		
P2	20 kΩ, 0.5% linear, 10-turn wire wound potentiometer		
P3	250 kΩ, 10%, 2 W carbon potentiometer		
FS	Function switch, ceramic rotary, 5P5T		
GS	Gain switch, ceramic rotary, 6P5T		
MS	Meter switch, ceramic rotary, SP6T		
M	0 to 50 µA meter, Lafayette Radio No. 99-R-5042		

conductance. Diode D_2 is used to provide a dc readout. It is easy to show that the height difference between the arms of the manometer is directly proportional to pressure and that the cell conductance is directly proportional to the height that the electrodes are immersed in the fluid. Thus a dc output signal is obtained which is directly proportional to the pressure change in the system. The linearity of response of the instrument was verified experimentally and was reproducible within a few parts per thousand for full scale pressure ranges from 0 to 6.30 and 0 to 0.315 mm of mercury, gauge (real pressures corrected for volume change). These ranges correspond to changes in the number of micromoles of gas in this particular system of 0 to 280 and 0 to 14.0, respectively. The reactors used were similar to ones to be described later.

In recent years, instruments which have frequently been called "electronic manometers" have become commercially available [7]. These devices do not employ a manometer in the classical sense. Rather, the pressure input signal is transformed into an electronic output signal. Resistive, inductive, capacitive, and piezoelectric circuit elements have all been used for this purpose. Curran and Driscoll [1] have described the use of a capacitive type pressure transducer system for pressuremetric titrations. A functional block diagram of the Datametrics system (Datametrics Division of C.G.S. Scientific, Waltham, Mass.) is shown in Fig. 5. Port P1 of the Type 511 Barocel transducer is connected to the working reactor and port P2 to a reference reactor. A pressure difference between these two ports results in deflection of a membrane diaphragm separating them. Physically, the diaphragm is the common plate between two parallel plate capacitors. Electronically,

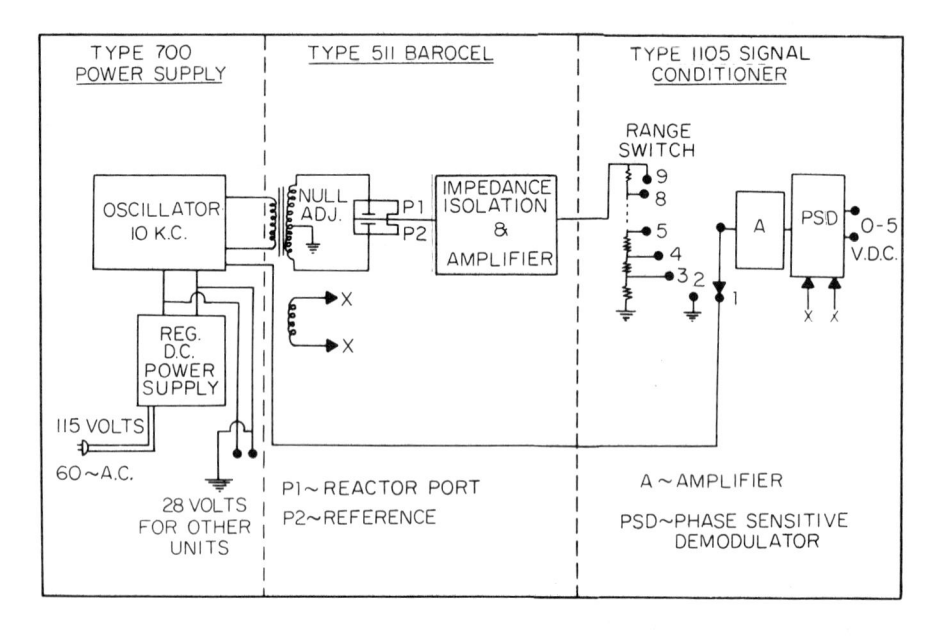

FIG. 5. FUNCTIONAL BLOCK DIAGRAM OF THE DATAMETRICS PRESSURE TRANSDUCER SYSTEM. FROM REF. [1] WITH PERMISSION.

2. UNUSUAL END-POINT DETECTION METHODS

the two capacitors are arranged as an ac voltage divider and placed as two arms of an LC bridge circuit. The effect of diaphragm displacement is to amplitude modulate the 10-kHz carrier signal supplied to the bridge by the Type 700 power supply. Following a unity gain impedance isolation amplifier and an output transformer, the modulated carrier signal is presented to the Type 1105 Signal Conditioner where circuits are provided for attenuation, amplification, and synchronous phase sensitive demodulation. The result is a 0 to 5 V dc signal, proportional to pressure, for each full scale setting of the range switch (attenuator). The 511 Barocel is available in standard ranges of 0 to 1, 0 to 10, 0 to 100, and 0 to 1000 Torr and 0 to 1, 0 to 10, and 0 to 100 psi, gauge, differential, absolute, and sealed absolute. The attenuator settings are X1, X0.3, X0.1, X0.03, X0.01, X0.003, and X0.001 of the nominal full scale rating of the transducer. System sensitivity is 1 part in 50,000 parts. Properly operated, the system has a response that is precise, accurate, and linear to within a few parts per thousand.

Curran and Swarin (13) have described the use of an unusual pressure transducer for pressure measurements. The base-emitter junction of some transistors exhibits the piezoelectric effect; that is, the junction is stress sensitive and a mechanical input will modulate the electrical output of the transistor. A unit called the Pitran (Stow Laboratories, Inc., Hudson, Mass.) is commercially available in the form of a silicon planar NPN transistor mounted in a TO-46 can. The top of the can acts as a force summing diaphragm and a stylus transmits the force to the base-emitter junction. The underside of the diaphragm is

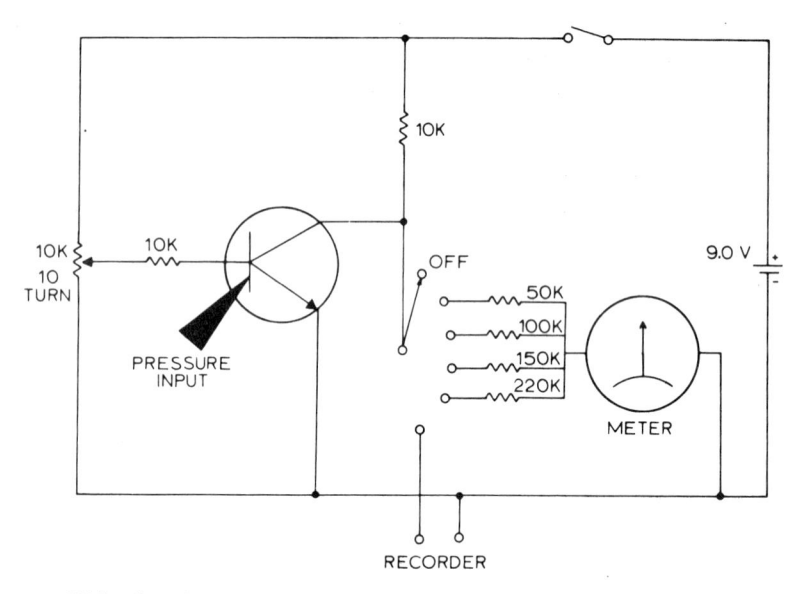

FIG. 6. CIRCUIT DIAGRAM OF THE PITRAN PRESSURE TRANSDUCER SYSTEM. FROM REF. [13] WITH PERMISSION.

exposed to atmospheric or a reference pressure through holes in the physical base of the transistor. This arrangement produces a differential pressure transducer. Figure 6 shows the circuit digram for the electrical arrangement of the transducer. Power was supplied to this common emitter amplifier circuit with a 9 V transistor battery and all components except the Pitran were mounted in a 5-1/4 x 3 x 2-1/8-in. Minibox. The Pitran used in this work was a Model PT-M2 with a nominal pressure range of 0.25 psid. Figure 7 presents an experimentally determined transfer function for the unit used. The differential nature of the device is clearly indicated. The linear output region found corresponded to 0.39 psid. This loss in sensitivity was attributed to lateral stresses induced in the Pitran by the method used to mount it (Fig. 8). It is also noteworthy that the collector-

2. UNUSUAL END-POINT DETECTION METHODS

emitter output voltage is not zero for zero Δp across the diaphragm. When working at high sensitivity or with pressures which are always unidirectional with respect to the reference pressure, it is convenient to null out the initial V_{CE}. As shown in Fig. 8, the Pitran was epoxied into the ball section of a ball and socket joint and a transistor socket was placed in the socket section of the joint. Ball joints E and F made connections to the reference and working reactors, respectively. Coulometric generation of titrant was employed using reactors to be described below. This transducer system is simple, inexpensive, portable, and capable of yielding titration results of very good precision and accuracy. It lacks versatility in that a given Pitran is not capable of multi-

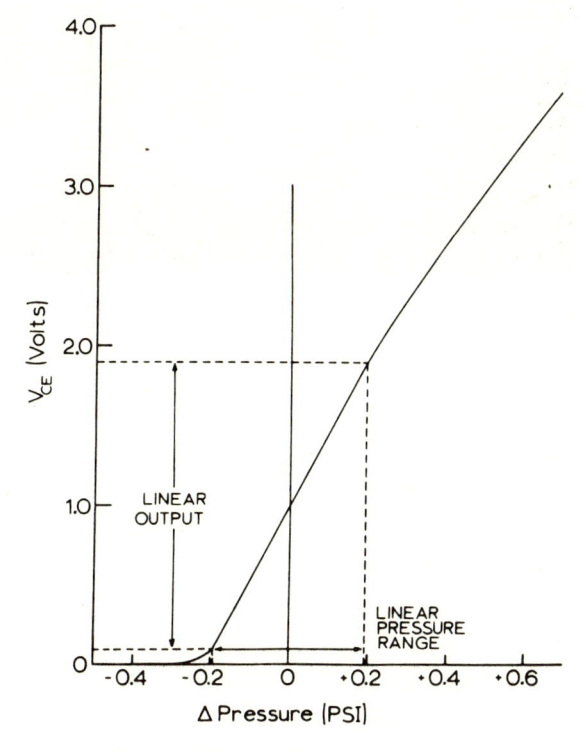

FIG. 7. INPUT PRESSURE-OUTPUT VOLTAGE DIAGRAM FOR PITRAN PRESSURE TRANSDUCER SYSTEM. FROM REF. [13] WITH PERMISSION.

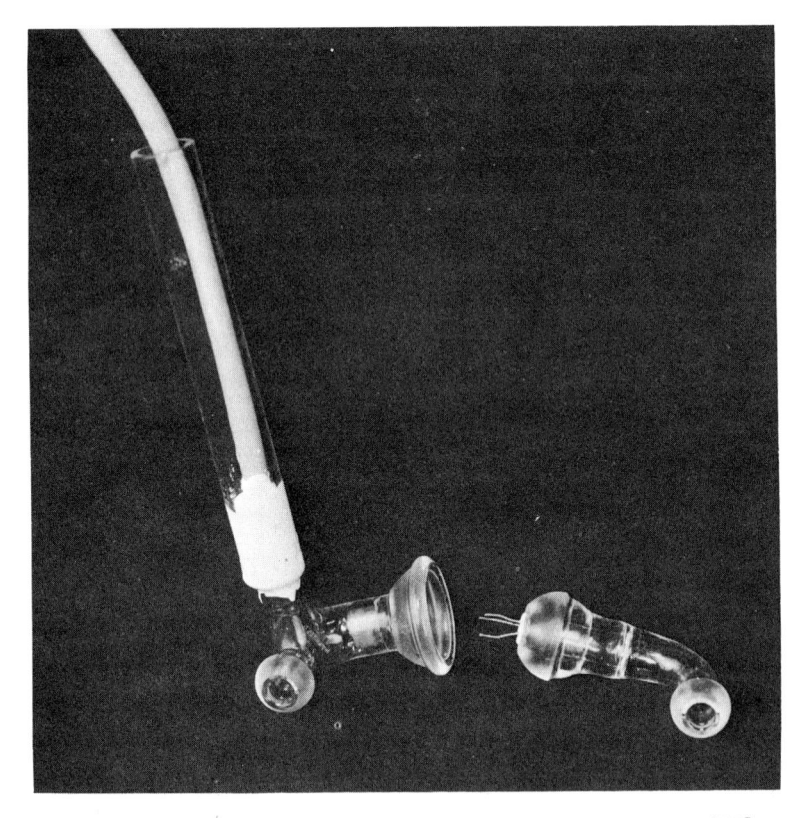

FIG. 8. PHOTOGRAPH OF MOUNTED PITRAN. FROM REF. [13] WITH PERMISSION.

range operation but this is partially overcome by the availability of nine different models with ranges from 0.1 to 20 psid.

The pressure transducer systems described by Curran and co-workers have been used with volumetric and/or coulometric generation of titrant. A typical reactor design for volumetric work is shown in Fig. 9. Microburet A is used for titrant delivery and has a bent tip so titrant may be inserted into the titrate solution when the latter is being very vigorously stirred. Microburet B is used for volume compensation. A

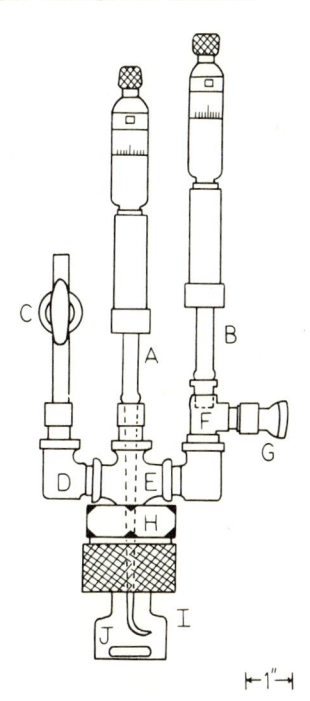

FIG. 9. TYPICAL REACTOR FOR VOLUMETRIC ADDITION OF TITRANT. A, MICROBURET FOR TITRANT; B, MICROBURET FOR VOLUME COMPENSATION; C, 2-mm STOPCOCK CEMENTED IN 1/4-IN. STAINLESS-STEEL BUSHING; D, 1/4-IN. STAINLESS-STEEL ELBOW: E, 1/4-IN. STAINLESS-STEEL CROSS JOINT; F, BRASS COMPRESSION TEE JOINT FOR 1/8-IN. PIPE; G, CONNECTION TO TRANSDUCER, $ 12/9 GLASS SOCKET, WITH 1/2-IN. STEM CEMENTED INTO 1/8-IN. BUSHING; H, STAINLESS-STEEL GLASS PIPE COUPLING FOR 3/4-IN. GLASS PIPE IN WHICH E IS CEMENTED; I, SOLUTION CONTAINER; J, TEFLON COATED MAGNETIC STIRRING BAR.

small plug of water is maintained in the barrel of this buret to help preserve a gas tight seal. If, for example, 10.00 µl of titrant are delivered with microburet A, microburet B is then backed off by exactly the same volume. This procedure is necessary because at high transducer sensitivities, the compression of the gas phase produced by the addition of titrant is more than sufficient to produce a full scale pressure change in the system. Stopcock C is used to admit gas to the system if desired. A similar vessel is used on the reference side

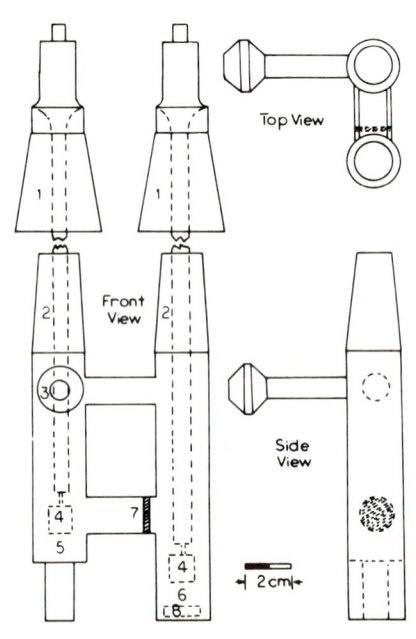

FIG. 10. PRESSUREMETRIC TITRATION REACTOR FOR COULOMETRIC GENERATION OF TITRANT. 1, ℥ 29/42 OUTER MEMBER GLASS JOINT; 2, ℥ 29/42 INNER MEMBER GLASS JOINT; 3, ℥ 18/5 BALL JOINT; 4, 2-cm^2 PLATINUM ELECTRODES; 5, CATHODE COMPARTMENT; 6, ANODE COMPARTMENT; 7, 10-mm FINE POROSITY FRIT; 8, MAGNETIC STIRRING BAR.

and both are immersed in a constant temperature bath. Bath temperature control to ±0.001°C is desirable when titrating solutions in the 10^{-4} to 10^{-5} M concentration range. The Barocel transducer described earlier houses some solid-state circuitry and a small amount of heat is produced in the unit. A Lucite housing with a water tight aluminum lid was constructed so the Barocel could also be immersed in a bath.

A diagram of the reactor used by Curran and Curley [4] for coulometric generation of titrant is shown in Fig. 10. This is a slightly modified electrolytic H-cell. An additional cross arm between the anode and cathode compartments allows gases generated in either compartment to have access to the rest of the reactor. The reference vessel was identical in

2. UNUSUAL END-POINT DETECTION METHODS   113

all respects to the one shown in Fig. 10 except that it was a mirror image. Connections to the pressure transducer system were made at the ball and socket joints. Platinum wire connections to the flag electrodes were sealed in glass tubing which was ring sealed to the female section of the standard taper joint. Electrical connection to the electrodes was made through mercury pools in the bottom of the tubing and copper or nichrome wire leads inserted through the top. The entire reactor, except for the top of the glass tubing containing the electrical leads, was submersed in the constant temperature bath. Stirring was done magnetically using submersible stirring motors.

All of the pressure transducer systems mentioned previously that have a transducer to convert a pressure signal to an electrical one have a dc output voltage so any dc voltage measuring device having a range compatible with the transducer system may be used for readout. A strip chart recorder is convenient since it is easy to establish when equilibrium has been reached from the strip chart readout. Higher precision and accuracy can be achieved by using potentiometers or digital voltmeters, and their use is justified with these pressure transducer systems.

D. Applications

The reactions of hydrazine are useful in pressuremetric work since they can yield nitrogen gas. Curran and Driscoll [1] used the iodometric titration of iodate with hydrazine to test their equation for the pressuremetric titration curve with volumetric addition of titrant. Curve A in Fig. 11 shows very good agreement between experimental and theoretical titration curves for the titration of 5.00 ml of 1.785×10^{-2} M

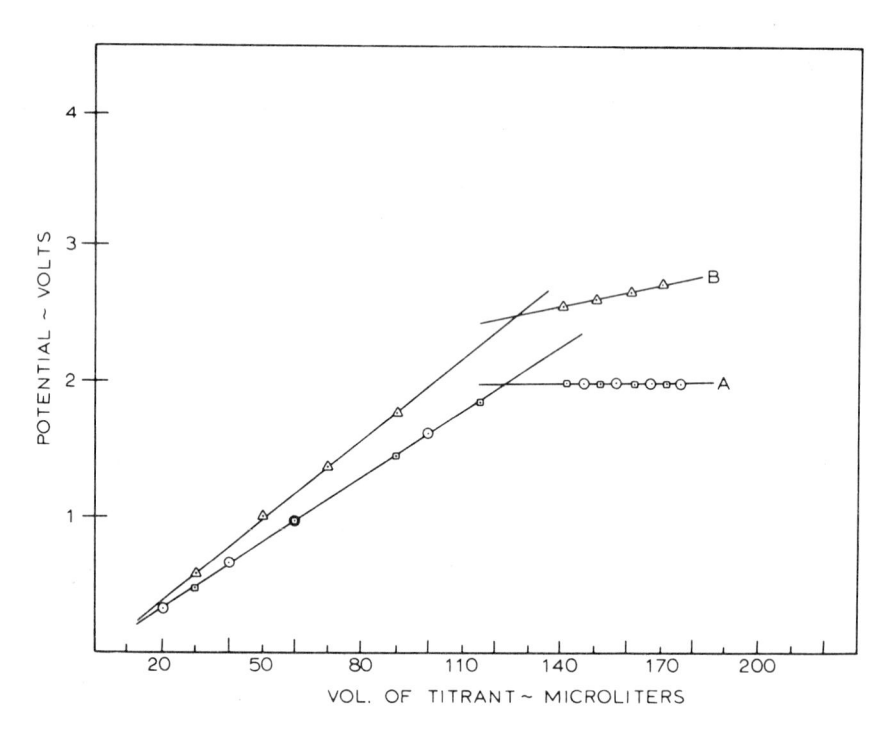

FIG. 11. PRESSUREMETRIC TITRATION CURVES. CALCULATED (○) AND EXPERIMENTAL (□) CURVES FOR THE TITRATION OF 5.00 ml OF 1.785×10^{-2} M POTASSIUM IODATE SOLUTION WITH 1.078 M HYDRAZINE SULFATE-M-X0.3 RANGE SETTING. EXPERIMENTAL (▲) CURVE FOR THE TITRATION OF 5.00 ml OF 1.785×10^{-3} M POTASSIUM IODATE SOLUTION WITH 0.1068 M HYDRAZINE SULFATE-M-X0.03 RANGE SETTING. FROM REF. [1], WITH PERMISSION.

KIO_3 with 1.078 M hydrazine sulfate. Nitrogen is produced by the following reactions:

$$6H^+ + IO_3^- + 5I^- \rightleftharpoons 3I_2 + 3H_2O \qquad (21)$$

$$2I_2 + N_2H_4 \cdot H_2SO_4 \rightleftharpoons N_2 + SO_4^{-2} + 6H^+ + 4I^- \qquad (22)$$

Correlation analysis on the least squares lines for a similar

2. UNUSUAL END-POINT DETECTION METHODS

titration curve produced correlation coefficients of 0.982 and 0.999 for the lines before and after the end point, respectively, showing that the data fit a linear relationship very well. The precision and accuracy of the titrations were excellent, being a few parts per thousand. Curve B of Fig. 11 illustrates the compression effect produced by titrant addition without volume compensation. In this case, the pressure signal due to gas evolution is much larger than that due to compression and the effect of the latter is to increase the slope of both lines. However, it is possible to dilute the iodate sample enough to produce the reverse situation where the compression signal is much larger than the gas evolution signal. Here the compensation technique becomes mandatory. The rate of gas evolution depends very much on the rate of stirring. With very vigorous stirring, an equilibrium pressure signal can be achieved within 1 or 2 min. Driscoll [2] has titrated as little as 256 µg of iodine with hydrazine sulfate in a phosphate buffer and found an accuracy of +0.5% (relative) and a precision of ±1.2% (standard deviation). Gottlieb [14] used a bicarbonate buffer for the same reaction. Before the end point both nitrogen and carbon dioxide are evolved, while after the end point only the CO_2 comes off. Using visual observation of the change in the manometer liquid level as the end-point technique, he was able to titrate about a tenth of a gram of iodine with a precision and accuracy of a few parts per thousand.

Table 1 summarizes applications for both volumetric and coulometric addition of titrant. Reactions from redox and acid-base chemistry are included. An interesting example is the titration of oxyhemoglobin with ferricyanide [2]. Earlier work on this reaction using potentiometric end-point detection had shown a very slow approach to equilibrium potentials [15]. It seemed unreasonable that this was a matter of slow attainment

TABLE 1

Applications of Pressuremetric Titrations

Titrant	Titrate	Comments	Reference
\multicolumn{4}{l}{A. Volumetric Titrant Addition}
NaOH	HCl	Visual ep, absorption of CO_2	8
Na_2CO_3	H_2SO_4	Visual ep, CO_2 evolution before the ep, CO_2 absorption after the ep	8
HCl	NaOH	Visual ep, CO_2 evolution	8
HCl	Na_2CO_3	Visual ep, no gas evolution to the HCO_3^- ep, CO_2 evolution to the 2nd ep	8
N_2H_4	I_2	HCO_3^- buffer, visual ep, before the ep N_2 and CO_2 evolve, after ep only CO_2	14
As_2O_3	I_2	HCO_3^- buffer, visual ep, CO_2 evolution	14
N_2H_4	IO_3^-	Visual ep, N_2 evolved before ep, pH on the acid side	14

2. UNUSUAL END-POINT DETECTION METHODS

As$_2$O$_3$, presence of excess I$^-$	IO$_4^-$	Visual ep, liberated I$_2$ titrated with the As$_2$O$_3$, CO$_2$ evolved from HCO$_3^-$ buffer	14
N$_3^-$	Ce(IV)	Graphical ep, N$_2$ evolution	9
Ce(IV)	Fe(II)	Graphical ep, N$_2$H$_4$ added to serve as indicator, reverse L-shape curve	9
ClO$^-$	NH$_4^+$	Graphical ep, N$_2$ evolved, excess bromide present so BrO$^-$ is the titrant	9
ClO$^-$	N$_2$H$_4$	Graphical ep, N$_2$ evolved, excess bromide present so BrO$^-$ is the titrant	9
NO$_2^-$	H$_2$NSO$_3$H	Graphical ep, N$_2$ evolved	9
HCl	NaOH	HCO$_3^-$ added as indicator, reverse L-shape titration curve	9
HCl	Na$_2$CO$_3$	Graphical ep, both ep's available	9
Ce(IV)	NaN$_3$ and N$_2$H$_4$	Mixture analysis, graphical ep	9
Cr$_2$O$_7^{2-}$	Fe(II)	Excess N$_2$H$_4$ added to serve as indicator, N$_2$ evolution after the ep	17

Table 1--Continued

Titrant	Titrate	Comments	Reference
$Cr_2O_7^{2-}$	Sn(II)	Excess N_2H_4 added to serve as indicator, N_2 evolution after the ep	17
Ag^+	N_2H_4	Basic solution, inverted L-shape curve	10
Ag^+	I^-	Basic solution, N_2H_4 added as indicator, reverse L-shape curve	10
Ag^+	CN^-	Basic solution, N_2H_4 added as indicator, reverse L-shape curve	10
N_2H_4	IO_3^-	Phosphate buffer, inverted L-shape curve	1, 11
MnO_4^-	H_2O_2	O_2 evolution, graphical ep	2
Ce(IV)	H_2O_2	O_2 evolution, graphical ep	2
Ce(IV)	N_3^-	N_2 evolution, graphical ep	2
IO_3^-	N_2H_4	N_2 evolution, graphical ep	2
I_2	N_2H_4	N_2 evolution, graphical ep	2

2. UNUSUAL END-POINT DETECTION METHODS

Fe(CN)$_6^{3-}$	Oxyhemoglobin	O$_2$ evolution, graphical ep	2
XeO$_3$	Fe(II)	Xe evolution, graphical ep	16
XeO$_3$	Ti(III)	Xe evolution, graphical ep	16
	B. Coulometric Titrant Addition		
BrO$^-$	NH$_4^+$	N$_2$ evolved, graphical ep	11, 13
Br$_2$	Carboxylic acid hydrazides	N$_2$ evolved from the titration reaction, H$_2$ evolved from the counter electrode reaction, graphical ep	4

of equilibrium of the chemical reaction, but rather slow
attainment of the equilibrium electrode potential. The pressuremetric study verified that the chemical reaction was fast
and titrations were performed successfully on 50 to 60 mg
samples of oxyhemoglobin.

Xenon trioxide has been relatively little studied as an
oxidizing agent in titrimetry. The principal reason for this
is the lack of a suitable method of end-point detection for
direct titration reactions. Potentiometric end points are
difficult since the xenon half reaction is irreversible.
Spectrophotometric methods are also marginal since the only
absorption band appears on the far edge of the near uv. Pressuremetric end-point detection is well suited to the task
since xenon trioxide always releases xenon gas. Ideally,
there is a six electron change according to the half reaction:

$$XeO_3 + 6H^+ + 6e^- \longrightarrow Xe + 3H_2O \qquad (23)$$

Blanchette [16] has studied the titration of Fe(II) and of
Ti(III) with this oxidant using the pressuremetric technique.
Sulfuric acid solutions were used exclusively. Both the iron
and titanium systems exhibited a stoichiometry of slightly
less than 6 moles of reductant per mole of xenon trioxide.
Proof was found for the production of oxygen as a side product
in both of these reactions and some peroxide is apparently
also formed in the titanium reaction. By standardizing techniques and empirically calibrating the procedure, highly
accurate, precise and sensitive titrations of both Fe(II) and
Ti(III) were performed. Twenty milliliter samples containing
1.7 to 4.2 mg of Ti(III) were analyzed with a precision and
accuracy of a few parts per thousand and similar results were
found for Fe(II) solutions containing 0.5 to 4 mg of iron.

Gottlieb has demonstrated the pressuremetric titration of
mixtures which contain two substances which release gas [9].
Equal volumes (2.50 ml) of 0.1 N hydrazine sulfate and 0.1 N

2. UNUSUAL END-POINT DETECTION METHODS

sodium azide were titrated with Ce(IV). Two distinct changes in slope were seen in the titration curves. The first end point was poor but the second was accurate to within a few parts per thousand. Similar results were obtained for a mixture of hydrazine sulfate and ammonium sulfate titrated with hypochlorite in a solution containing bromide. In another vein, Gottlieb has used a gas evolution reaction to detect the end point for another reaction which does not release gas. Examples include the titration of Fe(II) with $Cr_2O_7^{2-}$ in the rresence of hydrazine [17], the titration of arsenite with hypochlorite in the presence of ammonium ion, and the titration of sodium hydroxide with hydrochloric acid in the presence of bicarbonate [9]. The titration curves are reverse L-shaped.

Curran and Curley [4] combined pressuremetric end-point detection with coulometric generation of titrant for the analysis of acid hydrazides. The reaction can be expressed as

$$2Br_2 + R\text{-}\underset{\overset{\|}{O}}{C}NHNH_2 + H_2O \longrightarrow R\text{-}\underset{\overset{\|}{O}}{C}\text{-}OH + N_2 + 4HBr \qquad (24)$$

where the bromine is generated at a platinum anode. Not all of the acid hydrazides studied were found to obey Eq. (24) but 6 to 10 mg samples of isonicotinic-, p-toluic-, and phenylacetic acid hydrazides were titrated with very good precision and accuracy. Hydrogen gas was generated at the cathode of the electrochemical cell and nitrogen by the chemical reaction. The titration curve shows a well-defined change in slope at the end point. Carbohydrazide could be titrated with excellent precision and in good agreement with an amperometric end point but either end point indicated a bromine to carbohydrazide ratio of 3.60/1. Speculation by previous workers [18] that the solid carbohydrazide existed as the hemihydrate could not be proven using differential scanning calorimetry, thermogravimetric analysis, or elemental analysis. Thus, the bromine oxidation of carbohydrazide does not proceed according to Eq. (24) and the actual reaction remains unclear. Several dihydrazides were also titrated but only one successfully-- adipic dihydrazide. This compound behaves in the manner expected for a bifunctional acid hydrazide and yields nitrogen accordingly. However, malonic and succinic dihydrazides did not produce nitrogen as expected. Amperometric titration of the same solutions indicated bromine uptake but pressuremetric measurements showed little or no nitrogen evolution. The difficulty here is probably one of both reaction rate and stoichiometry, and the bromine oxidation of acid hydrazides as expressed by Eq. (24) cannot be taken as a general method for this class of compounds. Excellent results can be obtained in specific cases and pressuremetric end points provide data with high precision and accuracy.

Curran and Swarin [11,13] coulometrically generated hypobromite from bromine in basic solution to titrate ammonium ion using both the electronic manometer and the Pitran pressure transducer systems described in the apparatus section. With the former instrument, 0.5 mg of NH_4^+ could be titrated with a precision of ±0.5% and an accuracy of 0.2%. Using the Pitran system, approximately 0.9 mg was determined with a relative standard deviation of ±0.4% and a relative accuracy of ±0.1%. The amounts of materials commonly produced by coulometric generation are compatible with the pressuremetric technique. Further, coulometric generation of titrant avoids the compression effect associated with volumetric addition of titrant and its use is recommended wherever possible.

2. UNUSUAL END-POINT DETECTION METHODS

III. CRYOSCOPIC TITRATIONS

A. Introduction

Cryoscopic titrations were invented by Bruckenstein and Vanderborgh and first reported in a paper in 1966 [19]. Anyone who has driven on the highway in a winter storm in cold climates is familar with the fundamental phenomenon underlying the method. Ice and snow on roads is melted by application of a salt since the freezing point of a dilute salt solution is lower than the freezing point of water. It is not necessary that an electrolyte be used for this purpose but it is more efficient to do so since the lowering of the freezing point is a colligative property of solutions and therefore depends only on the total number of solute particles present.

B. Theory and Methods

Consider a dilute solution of a nonelectrolyte contained in the two phase system, solution and frozen solvent. The solute undergoes no dissociation or association reactions. The following equation for the lowering of the freezing point is well established and its derivation can be found in nearly any textbook on physical chemistry:

$$T_0 - T = (\Delta T_f)_0 = (RT_0^2/\Delta H_{fusion})(n_2/n_1 + n_2) \quad (25)$$

where T_0 is the freezing point of the pure solvent, T is the freezing point of the solution containing n_2 moles of solute and n_1 moles of solvent, ΔH_{fusion} is the heat of fusion of the solvent in calories per mole, and R is the gas constant taken in calories degree^{-1} mole^{-1}. Since a temperature difference appears on the left-hand side of Eq. (25), either Centigrade or Kelvin degrees may be used but the latter must be used on the right-hand side. The assumptions involved in the derivation of this equation are: (a) Raoult's law applies. (b) The solvent freezes as a pure solid. (c) The heat of fusion is constant between temperatures T_o and T. (d) The solvent vapor obeys the ideal gas law. (e) The solution is dilute enough that the quantity, $-\ln(1-N_2)$ is equal to N_2, the mole fraction of solute. (f) The solution is dilute enough that T_o and T differ only a little so that the product $T_o T$ is given by T_o^2. Since the solution must be dilute, n_2 is small compared with n_1 and the sum, $n_1 + n_2$, can be taken as simply equal to n_1. Then

$$(\Delta T_f)_o = (RT_o^2/\Delta H_{fusion})(n_2/n_1) \qquad (26)$$

The number of moles of solute is equal to the product of the volume of solution, V, in liters times the equilibrium solute concentration, [C], in moles per liter. We have

$$(\Delta T_f)_o = RT_o^2 V[C]/\Delta H_{fusion} n_1 \qquad (27)$$

All of the terms on the right-hand side of Eq. (27) are sensibly constant except [C] and we define the molar freezing point constant, K_f, as

$$K_f = RT_o^2 V/\Delta H_{fusion} n_1 \qquad (28)$$

Therefore

$$T_o - T = (\Delta T_f)_o = K_f C \qquad (29)$$

Equation (29) is fundamental to cryoscopic titrations but it needs modification for solutions of electrolytes. Since a strong electrolyte will dissociate into ions when dissolved in water, it might be expected that the right-hand side of Eq. (29) should be multiplied by a small whole number equal to the number of ions indicated by the ionic or molecular formula of the material. This is true for solutions of electrolytes at infinite dilution and approached for dilute solutions. The van't Hoff i factor provides a quantitative expression of this effect. Let ΔT_f be the freezing point lowering for a solution of electrolyte of concentration C moles per liter and let $(\Delta T_f)_o$ be the freezing point lowering for a solution of an

2. UNUSUAL END-POINT DETECTION METHODS

ideal nonelectrolyte of concentration C mole per liter. The van't Hoff i factor is defined as

$$i = \Delta T_f / (\Delta T_f)_o \qquad (30)$$

Substituting from Eq. (29) into Eq. (30) and rearranging, we have

$$\Delta T_f = iK_f[C] \qquad (31)$$

The freezing point lowering for an electrolyte is greater than that for the same concentration of nonelectrolyte by the factor i. For very dilute solutions i is about equal to the number of ions expected from the ionic or molecular formula of the electrolyte but as the concentration increases, i decreases.

In view of Eqs. (29) and (31), the freezing point lowering is proportional to the total solute concentration in solution and it is this fact which has been exploited by Bruckenstein and Vanderborgh in using freezing point measurements to locate titration end points. The method requires that the change in the total concentration of solute particles before the end point be different from the change in the total concentration of solute particles after the end point. In many cases, this leads to linear-segmented type titration curves but only a distinct change in slope is necessary to detect an end point. Consider the neutralization titration in water of hydrochloric acid with sodium hydroxide. No change in the freezing point is predicted prior to the end point because the hydronium ion is merely being replaced in solution by the cation of the base.

$$Na^+ + OH^- + H^+ + Cl^- \rightleftharpoons H_2O + Na^+ + Cl^- \qquad (32)$$

Beyond the end point, there is an increase in the number of solute particles present due to the cation and the anion of the base added in excess and a decrease in the freezing point of the solution is expected. Operationally, concentrated titrant is added by a piston driven constant rate buret and the equilibrium freezing point is measured with a strip chart recorder as the unbalance signal from a Wheatstone bridge circuit using a thermistor as one arm of the bridge. The titration curve is continuously plotted on the recorder which has a time axis proportional to the volume of titrant added.

With the strip chart recording, the end-point datum is in hand but alternative ways of treating the data are useful. Let T_{in} be the initial equilibrium temperature of the solution to be titrated and T_e be the equilibrium temperature at any time during the titration. We define ΔT as

$$\Delta T = T_{in} - T_e \tag{33}$$

Adding and subtracting T_o, the freezing point of the pure solvent, from the right-hand side of Eq. (33) and collecting terms

$$\Delta T = (T_o - T_e) + (T_{in} - T_e) \tag{34}$$

From Eq. (31), $(T_o - T_e) = i[C_e]K_f$ and $(T_{in} - T_o) = -i[C_{in}]K_f$. Substituting in Eq. (33) and dividing both sides by $[C_{in}]$

$$\Delta T/[C_{in}] = iK_f[C_e]/[C_{in}] - iK_f[C_{in}]/[C_{in}] \tag{35}$$

But, $[C_e]/[C_{in}] = (1 - \phi)$, where ϕ is the fraction titrated. Making the substitution and rearranging we have:

$$i\phi = \Delta T/K_f[C_{in}] \tag{36}$$

Equation (36) also describes a titration curve. Note that ϕ is allowed to have values greater than one. If the RHS is plotted versus ϕ, and i is constant, a straight line should result with a slop of i. Different slopes will be found before and after the end point. It sould be mentioned that the point of reference here for i is the initial solution. That is, at the start, $\Delta T = 0$, $\phi = 0$, and therefore, $i\phi = 0$. For solutions sufficiently dilute, the change in i is given by the change in the number of solute particles in solution. In the HCl-NaOH titration, there is no change prior to the end point and the slope of the titration curve is zero. After the end point, two additional solute particles are introduced and the slope is two. The advantage gained in plotting the titration curve this way is that

2. UNUSUAL END-POINT DETECTION METHODS

the slopes of the titration curve are related in a simple way to the stoichiometry of the reaction. The values of i for more concentrated solutions, while less than the number for dilute solutions, are reasonably identical for the same concentration of different electrolytes of the same charge type. This tends to indicate that the ionic strength of the solution rather than the concentration of individual solution particles is the prime factor in determing the value of i. Thus, linear titration curves can be expected even in the presence of a large concentration of innocuous electrolyte but the advantage mentioned above is lost.

The concept of i can be extended to include any process affecting the number of solute particles in solution such as dissociation, association of ion pairs, etc. The following treatment gives another method for normalizing the data which may yield still additional information concerning the chemistry of the titration reaction. Recalling our original definition of i we write

$$i = (\Delta T_f)_{in}/(\Delta T_f)_{ref} \tag{37}$$

where $(\Delta T_f)_{in}$ is the freezing point lowering produced by a concentration $[C_{in}]$ of the substance to be titrated, and $(\Delta T_f)_{ref}$ is the freezing point lowering produced by an identical concentration of an ideal nonelectrolyte. Note that the substance to be titrated may also be an ideal nonelectrolyte in which case the value of i would be one at the start. Substituting in the denominator of Eq. (37) from Eq. (29) and multiplying the right-hand side by $[C_e]/[C_e]$ where $[C_e]$ is the equilibrium concentration of the sought-for-constituent at any time in the titration

$$i = (\Delta T_f)_{in}[C_e]/K_f[C_{in}][C_e] \tag{38}$$

Again, $[C_e]/[C_{in}] = (1 - \phi)$ and:

$$i = (\Delta T_f)_{in}(1 - \phi)/K_f[C_e] \tag{39}$$

But, $K_f[C_e] = (\Delta T_f)_e$ and therefore:

$$i = (1 - \phi)(\Delta T_f)_{in}/(\Delta T_f)_e \tag{40}$$

Equation (40) is valid up to and including the equivalence point. For values of ϕ greater than one, analogous considerations lead to a similar equation except that the term, $(1 - \phi)$ is replaced by $(\phi - 1)$. $(\Delta T_f)_e$ is related to the total number of solute particles present at any time during the titration while $(\Delta T_f)_{in}$ gives the initial condition. Both are experimentally measured. A plot of i vs ϕ will not only give the end point data, but it can provide information concerning the stoichiometry of the reaction, any possible stepwise reaction of titrant and titrate, and any association or dissociation processes of reactants and/or products.

C. Apparatus and Experimental Methods

The apparatus required for cyroscopic titrations is quite simple. The titration vessel is a stoppered wide mouth Dewar flask equipped with the delivery tip of a constant rate buret, a thermistor, and a means for stirring. In the work of Bruckenstein and Vanderborgh, a thermistor with a nominal resistance of 2000 Ω at 25°C was used. The thermistor in series with a 0 to 20 kΩ variable resistor formed one arm of an approximately equal arm Wheatstone bridge circuit (20 kΩ arms). The arm opposite the thermistor arm was variable to allow balancing the bridge prior to the start of a titration. The power supply for the bridge was a series combination of two 1.35 V mercury batteries and a 10 kΩ voltage divider. Bridge unbalance was measured with a 1 mV strip chart recorder. Upon calibration, the temperature sensitivity of this bridge was 4.63 mV/°C at an applied bridge voltage of 0.8 V. The smallest change in the number of particles that the instrument could detect was reported as 2.3×10^{-5} moles for water as the solvent and 9.6×10^{-6} moles in benzene, the solution volume being about 40 ml in both cases. The sensitivity was not limited by the temperature measurement but by noise in the signal attributed to pieces of the frozen solvent striking the thermistor. Some kind of mechanical shielding might reduce this problem and improve sensitivity.

Vigorous stirring is required to insure that the temperature measurement is an equilibrium one and a magnetic stirrer rotated at its maximum speed was used. If the system gains heat either from its surroundings or from internal processes, the net effect is to melt some of the frozen solvent. This dilutes the solution and produces a temperature rise which must be minimized. Heat effects caused by the addition of a small volume of concentrated titrant to a few millimoles of sought-for-constituent contained in a much larger volume of solution hardly affect the titration curve. The rate of heat exchange of the system with its surroundings was 9 cal/min as determined by Bruckenstein and Vanderborgh. This was not high enough to produce poorly defined end points but it is sufficient to affect the slope of the normalized titration curves since the usual charge in the Dewar was 25 to 40 ml of solution and 25 to 30 g of frozen solvent, and the minimum time to a titration end point was about 4 min. Bruckenstein and Vanderborgh mention that dilution corrections were made in slope calculations for normalized titration curves [19].

The use of operational amplifier circuits for the temperature measurement is possible. For the simple inverter amplifier configuration, the following relationship holds:

$$e_o = -e_{in}(R_f/R_{in}) \qquad (41)$$

where e_o is the amplifier output voltage, e_{in} is the input voltage, R_f is the value of the feedback resistor, and R_{in} is the value of the input resistor. If e_{in} and R_f are constant, and R_{in} is a thermistor with a response $R_T = K/T$, where T is the temperature in degrees centigrade, then

$$e_o = -e_{in}R_fT/K = K'T \qquad (42)$$

and the output signal is directly proportional to the

temperature. This equation is, of course, only valid over a limited temperature range. Vanderborgh and Spall [20] have published a circuit design using a pair of matched thermistors arranged in a summing amplifier configuration [21]. Both thermistors are powered by an identical voltage but one voltage is of opposite sign to the other. Here, the output signal is given by [20]

$$e_o = (-e_{in}R_f/R_{T1}R_{T2})(R_{T1} - R_{T2}) \qquad (43)$$

Again, if an equation of the form $R_T = K/T$ is valid and if e_{in} and R_f are held constant,

$$e_o = K'(T_2 - T_1) \qquad (44)$$

and the output voltage is directly proportional to the temperature difference between thermistors. Vanderborgh and Spall reported that temperature differences as small as 4×10^{-4} °C were easily measured and this compares very favorably with a sensitivity of about 1×10^{-3} °C reported by Bruckenstein and Vanderborgh for their bridge circuit when powered by 0.8 V. The advantage in using the operational amplifier circuit in cryoscopic titrations would be that the reference freezing point could be continuously monitored and continuously subtracted from the freezing point of the solution being titrated. A similar result could be achieved with a two-active-arm Wheatstone bridge circuit but the operational amplifier temperature difference circuit has the advantage of equal power dissipation in each thermistor.

D. Applications

The results of Bruckenstein and Vanderborgh are summarized in Tables 2 and 3 [19]. Acid-base reactions and a precipitation reaction were studied in water as the solvent. Recorded titration curves generally had linear segments--at least in the vicinity of the equivalence point and normalized

2. UNUSUAL END-POINT DETECTION METHODS 131

TABLE 2

Aqueous Cryoscopic Titrations[a]

Solute	5 M Titrant	Buret rate[b]	Millimoles taken[c]	Precision (%)	Accuracy (%)
HCl[d]	NaOH	0.09748	2.500	1.5	-0.25
HCl[d]	NaOH	0.04874	2.500	0.2	-1.2
HOAc[d]	NaOH	0.02440	2.500	1.1	0.4
HOAc[d,e] HCl	NaOH	0.04874	0.998 1.000	0.5 0.6	-0.9 0.6
HOAc[d,e] HCl	NaOH	0.02440	0.998 1.000	0.8 1.0	0.1 0.5
AgNO$_3$[d]	NaCl	0.04874	2.500	1.0	0.2

[a] Adapted from Ref. [19] with permission.
[b] Titrant delivery rate in milliliters per minute.
[c] Average volume of liquid phase = 35 ± 5 ml.
[d] Average result of three determinations.
[e] Mixture of acetic and hydrochloric acids.

plots of the data yielded the slopes expected from the stoichiometry of the reactions. The precision and accuracy indicated in Table 2 is about 1% relative or better. The titration of an equimolar mixture of hydrochloric and acetic acids is of interest. Two clearly defined end points were observed in the titration curve, and the results listed in Table 2 are quite satisfactory. A result not shown in the table confirmed the notion developed in the previous section that cryoscopic titrations should be possible in the presence of a large concentration of inert electrolyte. A titration of HCl with NaOH similar to those listed in Table 2 was successful in the presence of 1 M KCl. The dilution effect due to melting ice gave rise to considerable change in the freezing point temperature but the end point was reported as evident from the

TABLE 3

Cryoscopic Titrations of Acids and Amine Bases in the Solvent Benzene[a]

Amine	Acid	M	Millimoles taken[b]	Precision (%)	Accuracy (%)	m
Benzyl	Trifluoroacetic	4.95[c]	1.10	--	-0.9	--
Dodecyl	Trifluoroacetic	4.95[c]	1.49	--	2.6	--
Dodecyl	Trichloroacetic	2.95[d]	0.497	--	1.3	--
		2.95[c]	0.994	0.7	-0.7	30
		5.10[c]	1.19	--	0	--
		5.10[d]	1.19	--	-2.2	--
Hexyl	Trichloroacetic	2.95[c]	0.953	1.0	+0.7	--
Benzyl	Trichloroacetic	2.95[c]	1.10	--	-0.9	--
		2.95[d]	1.10	1.0	0	--
Cyclohexyl	Trichloroacetic	2.95[c]	0.939	0.4	-1.6	--
		2.95[d]	0.939	--	-0.6	--
Piperidine	Trichloroacetic	5.10[d]	0.962	0.7	-1.4	--
Dibenzyl	Trichloroacetic	5.10[d]	1.00	0.1	0	2.3

2. UNUSUAL END-POINT DETECTION METHODS 133

N,N-dimethylbenzyl	Trichloroacetic	2.95[c]	1.06	--	-0.9	1.4
		2.95[d]	1.05	1.0	-0.7	30
Dodecyl	Acetic	4.32[d]	1.21	1.0	-0.3	--
Hexyl	Acetic	4.32[d]	0.953	1.3	-0.7	5.3
Piperidine	Acetic	4.32[d]	0.962	--	0.5	1.1
N,N-diemthylbenzyl	Acetic	4.32[d]	1.05	--	0	--
Dodecyl and dibenzyl	Trichloroacetic	5.10[d]	0.595	--	--	--
			0.501	--	--	--
			1.096 (total)	--	-1.3	--
Dodecyl and N,N-dimethylbenzyl	Trichloroacetic	5.10[d]	0.595	4.4	-5.2	--
			1.045	3.0	3.9	--
			1.640 (total)	0.7	0.5	--

[a]Adapted from Ref. [19] with permission.
[b]Average volume of liquid phase = 40 ± 5 ml.
[c]Buret delivery rate of 0.04874 ml/min.
[d]Buret delivery rate of 0.02440 ml/min.

strip chart recording and corresponded to the calculated equivalence point apparently within 1%.

Bruckenstein and Saito [22] used differential vapor pressure and ir measurements in a study of acid-base reactions in benzene and proposed the following equilibria to describe acid-base interactions in this solvent:

$$B + HX \rightleftharpoons BH^+X^- \qquad (45)$$

$$BH^+X^- = (1/m)(BH^+X^-)_m \qquad (46)$$

$$HX + (1/m)(BH^+X^-)_m = (1/n)(BH^+HX_2^-)_n \qquad (47)$$

The cryoscopic titration method is also useful in the study of ion pairs and ion-pair aggregates such as these. The titration curves were not linear but had well-defined end points. Accuracy and precision, as shown in Table 3, were about 1-2%. Normalized plots were made according to Eq. (40) assuming 100% titrated corresponded to the BH^+X^- equivalence point. The slopes of these nonlinear plots gave information regarding the species present and the state of aggregation. The value found for m of Eq. (46) is listed in the last column of Table 3. These were estimated from the slopes of the titration curves in the vicinity of the end point. Naphthalene was used as the solute for the $(\Delta T_f)_{ref}$ measurements. In all cases the value of i found at a fraction titrated of zero was one proving that all of the bases titrated were monomeric in benzene. The acids were dimerized to a considerable extent. The equilibrium constants for the dimerization reaction [Eq. (48)] were estimated from cryoscopic measurements in benzene at 5.5°C

$$2RCOOH \rightleftharpoons (RCOOH)_2 \qquad (48)$$

as $1/7 \times 10^3$ for trichloroacetic acid and $1/1.9 \times 10^4$ for acetic acid. Each end point listed in Table 3 was well defined at the BH^+X^- equivalency but end points could not be measured at 200 or 300% titrated. However, the slopes of the lines gave evidence for the presence of ion pairs of the type $(BH^+HX_2^-)$ which formed ion aggregates. Beyond 200% titrated, the slope of the lines was still less than that predicted for the addition of excess acid. In the early part of the titration, the slope of the titration curves suggested the presence of a species with stoichiometry, $B_2 \cdot HX$. The evidence in Table 3 lead to the conclusions that association of the acid-base adduct is stronger for primary amines than secondary than tertiary for a given acid and that both acid strength and geometry of the adduct are important factors.

Bruckenstein [23] has also reported the application of cryoscopic titrations to complexation chemistry involving metal ions and EDTA but no data are available.

On the strength of the results reported thus far in the literature, it appears that cryoscopic titrations should be a powerful tool for analytical studies in solution chemistry. The end-point detection technique is quite universal, the apparatus inexpensive, and the methodology reasonably simple.

IV. PHASE TITRATIONS

A. Introduction

Phase titrations are based on the solubility relationships of solutions or mixtures. A third reagent which may be a pure liquid or a solution is added to the system to be determined until a visible response corresponding to a phase change occurs. This process has many of the features of a titration such as a clearly defined end-point signal and controlled addition and measurement of the volume of reagent added, but it lacks the concept of equivalence and reaction in the sense of being able to calculate the amount of unknown present from the volume and concentration of the titrant used, and the stoichiometry or reproducibility of the reaction. In phase titrations, it is necessary to prepare calibration curves to relate the volume of titrant used to the composition of the unknown. The method is not a new one, having been presented in detail by Bogin [24] in 1924. It was probably used occasionally before that. However, considerable activity has appeared in the recent literature. Higuchi and Connors [25] have classified phase titrations as defined here as solubility titrations, but the former name is widely used in the literature while the latter is not.

B. Discussion of Methods

1. *The Phase Rule and Phase Diagrams*

The majority of phase titrations have involved ternary systems where three pairs of liquids may be considered. One of the pairs is partially miscible and the remaining two are miscible in all proportions. A consideration of the phase rule shows the basis for these titrations. Writing the phase rule we have

$$F = C - P + 2 \qquad (49)$$

where F, the number of degrees of freedom, is the number of independent intensive variables, C is the number of components, and P is the number of phases. For a three-component system which undergoes no chemical reactions

$$F = 5 - P \qquad (50)$$

At constant temperature and pressure with one phase present, the system is invariant (F = 0) when the relative amounts of two of the components are fixed. These conditions prevail in a phase titration of a homogeneous binary mixture with the remaining component of the three-component system. An end point appears when a sufficient amount of the third component has been added to produce a two-phase system, and it is

2. UNUSUAL END-POINT DETECTION METHODS

signaled by the appearance of turbidity. At constant temperature and pressure with two phases present initially, the only degree of freedom is the concentration of one of the components in one of the phases. Advantage is taken of this fact to titrate a binary mixture to the disappearance of turbidity (clarification end point).

Isothermal phase diagrams of three-component systems are most commonly graphed on an equilateral triangle. As Findlay and Campbell [26] have pointed out, two different methods of constructing the triangle have been used. In one case the height of the triangle is taken as unit length. Pure liquids are represented by the apexes of the triangle. The coordinates of a point within the triangle are determined by perpendiculars from the point in question to the sides of the triangle, the fraction of one component being given by the perpendicular distance from the point to the side opposite the apex representing the component. In the second method, the sides of the triangle are assigned unit length with the apexes the same as before. The coordinates of a point inside the triangle can be determined by lines drawn parallel to the apexes. For a mixture containing fractions a, b, and c of components A, B, and C, respectively, a line through the point and parallel to the apex of the A component will intersect both sides of the triangle forming the apex. The a coordinate of either of these intersections is the fraction of A present at the point under consideration. Results for b and c are obtained in a similar fashion. Since the definition of unit length is arbitrary, either method can be used for the interpretation of a phase diagram when it is not known which method was used in constructing it. A correct choice may, however, result in convenient units.

A phase diagram typical of the systems under consideration is shown in Fig. 12. The curved line is a solubility curve

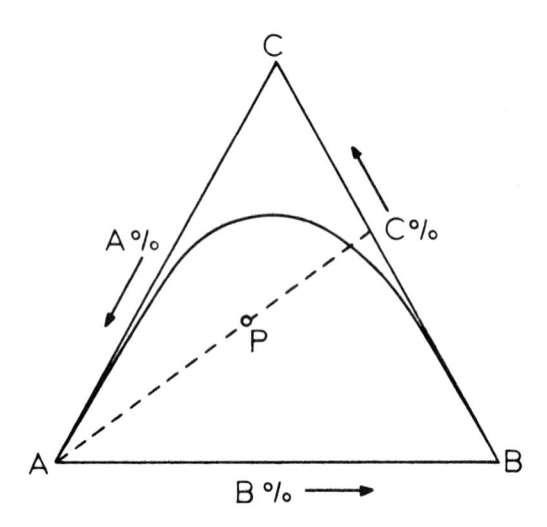

FIG. 12. REPRESENTATION OF A THREE-COMPONENT SYSTEM USING AN EQUILATERAL TRIANGLE. POINT P REPRESENTS THE TERNARY SOLUTION WITH THE COMPOSITION 40% OF A, 24% OF B, AND 36% OF C. FROM REF. [27] WITH PERMISSION.

called the binodal or miscibility boundary curve. Regions above the curve correspond to compositions of the ternary system where only one phase exists while regions below the curve correspond to compositions where two phases exist. The lines forming the sides of the triangle represent binary mixtures and the apexes pure liquids. In this diagram, A and B are immiscible in all proportions.

2. *Methods for Binary Mixtures*

The dotted line in Fig. 12, drawn from the A apex to a point on the opposite side represents the composition of all

2. UNUSUAL END-POINT DETECTION METHODS 139

possible mixtures of A, B, and C with a constant ratio of B to C of 40 to 60. A titration of the binary mixture, 40% B and 60% C with pure A will have one phase until enough A has been added to bring the ternary mixture to a composition represented by the intersection of the dotted line with the binodal curve at which point turbidity will appear. The upper part of Fig. 13 shows a series of such lines for 10/90, 20/80, 30/70, 40/60, and 50/50 mixtures of B to C (lines 1 through 5). The point of intersection of each line with the binodal curve represents a fixed volume of A added in each case and the lower part of

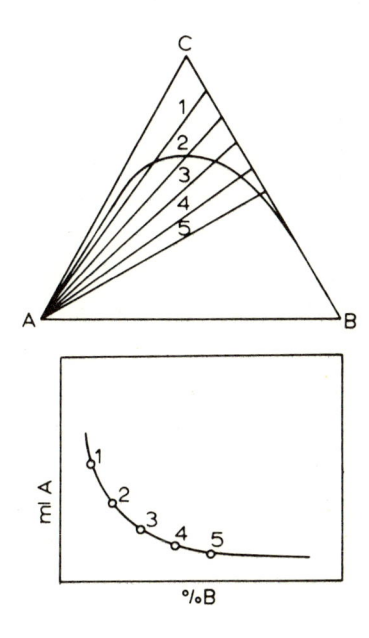

FIG. 13. THE RELATIONSHIP BETWEEN A THREE-COMPONENT PHASE DIAGRAM AND A CALIBRATION CURVE. EACH POINT ON THE CALIBRATION CURVE REPRESENTS THE INTERSECTION OF THE CORRESPONDING TITRATION LINE WITH THE BINODAL CURVE. FROM REF. [27] WITH PERMISSION.

Fig. 13 illustrates a calibration curve prepared by plotting these volumes versus the %B, which is the immiscible component. Rogers [27] has termed calibration curves of this nature, "Type A." The curve is nonlinear and not very useful in its extremities. An optimum range of %B exists where best results can be found. Rogers has defined this region as the range of %B requiring from about 1 to 10 ml of titrant for 10 ml samples. This should not be regarded as a rigorously binding definition but it offers a useful rule of thumb. Also, it is always possible to dilute mixtures of high per cent B or add B to mixtures of low per cent B to bring the composition of the sample into the optimum range. Further, Bogin [24] has shown that it is possible to shift the calibration curve along the horizontal axis by adding salts to the titrant since these change the solubility relationships involved.

Figure 14 illustrates a phase diagram with tie lines drawn. Recall that any ternary mixture whose composition falls under the binodal curve will separate into two phases. The composition of each of these phases is usually different and each is represented by a point on the binodal curve. The line connecting the two points is called the tie line. Each tie line must be determined experimentally. As we proceed up the diagram, the tie lines become shorter and shorter until the two points of intersection on the binodal curve merge at a single point called the plait point. The two phases now have identical compositions. Rogers and Ozsogomonyan [28] and Atwood [29] have discussed the factors affecting the quality of the end point obtained in phase titrations. In general, the best end points are found when the titration occurs along a line which is as perpendicular as possible to the binodal curve and to the tie line, and which passes as close to the plait point as possible. This arises for operational reasons. Titrations conducted along lines perpendicular to the tie line

2. UNUSUAL END-POINT DETECTION METHODS 141

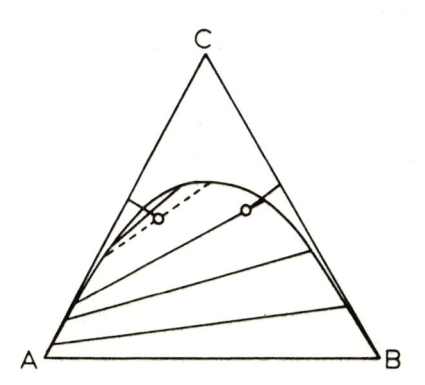

FIG. 14. A THREE-COMPONENT PHASE DIAGRAM SHOWING TIE LINES AND TITRATION LINES FOR THE TWO TITRANTS A AND B. THE LINE REPRESENTING TITRATION WITH A INTERSECTS THE SOLUBILITY CURVE WELL AWAY FROM THE PLAIT POINT GIVING A POOR END POINT, WHILE THE LINE REPRESENTING TITRATION WITH B INTERSECTS THE SOLUBILITY CURVE NEAR THE PLAIT POINT GIVING A GOOD END POINT. FROM REF. [28] WITH PERMISSION.

or close to the plait point promote the presence of sufficient volumes of both phases in the two-phase region to give sharply defined end points. Titrations approaching the binodal curve perpendicularly also tend to give sharp end points. The ability to see turbidity depends on the refractive indices of the phases and is affected by the degree of dispersion. Sometimes an end point can be improved by adding a chemical indicator to the system. Caley and Habboush [30] used iodine for this purpose. They titrated binary mixtures of aromatic hydrocarbons and the lower alcohols with water; the iodine indicator was particularly effective because of its characteristic violet color in the hydrocarbon phase. The amount of indicator used was too small to affect the solubility relationships involved so there was no blank correction. Spiridonova [31-35,37,38] and Spiridonova and Nikitin [36,39] have long advocated the use of organics as indicators in phase titrations. In much of their work they have used furfural for this purpose.

Spiridonova has also presented a way to treat the data which yields linear working curves in place of the Type A working curve [40]. This suggestion seems to have been widely ignored so a detailed example will be given here. Consider a binary mixture of A and B which is titrated with water. Let us say that B is an immiscible component of the system. A new variable called the degree of dilution, n, is used and defined as the ratio of the highest volume per cent of B to be encountered to the volume per cent of B for any other composition of the binary mixture which is more dilute in B. A plot of n versus the volume of water, V, required to reach the end point is a straight line over a considerable range of volume per cent B, and the linear region appears to include the most useful region of the Type A calibration curve. Caley and Habboush [30] have reported some Type A calibration curves with sufficient clarity that the data may be read from them. Table 4, columns 1 and 2, shows the data estimated from their calibration curve for the titration of benzene-absolute ethanol mixtures with water. The numbers in column 1 were obtained unambiguously but the volume of water was more difficult to ascertain. In the absence of the original data, there is of course some doubt about the correctness of the numbers in colume 2 of Table 4. However, they are probably close enough to being correct for the purpose here. Figure 15 is a plot of the data in columns 2 and 3 of Table 4 over the range of % benzene from 50 to 20. There is no question that the relationship is a linear one. The data point corresponding to n = 3.333 is not shown in the figure but it falls well above the line. The range of % benzene over which the curve is linear appears to be approximately from 45% (n = 1.111) to 20% (n = 2.500).

2. UNUSUAL END-POINT DETECTION METHODS 143

TABLE 4

Data and Calculations for the
Construction of a Linear Calibration Curve[a]

Benezene (%)	Volume of water (ml)	Degree of dilution (n)
50	5.0$_0$	1.000
45	6.0$_0$	1.111
40	7.1$_8$	1.250
35	8.4$_5$	1.429
30	10.2$_7$	1.667
25	12.7$_3$	2.000
20	16.1$_8$	2.500
15	20.6$_4$	3.333
10	28.3$_6$	5.000
5	43.6$_9$	10.000

[a]Data taken from a Type A calibration curve in Ref. [30].

This corresponds well with results reported by Spiridonova for titrations of ethanol-isoamylalcohol and dioxane-isoamylalcohol mixtures with water. In the former case, the working curve was linear from 40 to 15% isoamylalcohol (highest volume per cent = 50, n from 1.25 to 3.33) and in the latter case, from 60 to 15% (highest volume per cent = 60, n from 1.20 to 2.40) [40]. A linear calibration curve is much preferred over a nonlinear one and Spiridonova's method should be considered for adoption by many workers.

Dunnery and Atwood [41] have devised an experimental technique which directly provides a linear calibration curve. The basis for the method can be explained with the aid of the phase diagram shown in Fig. 16. For the binary pair, A and B, which are miscible in all proportions, a third component, C,

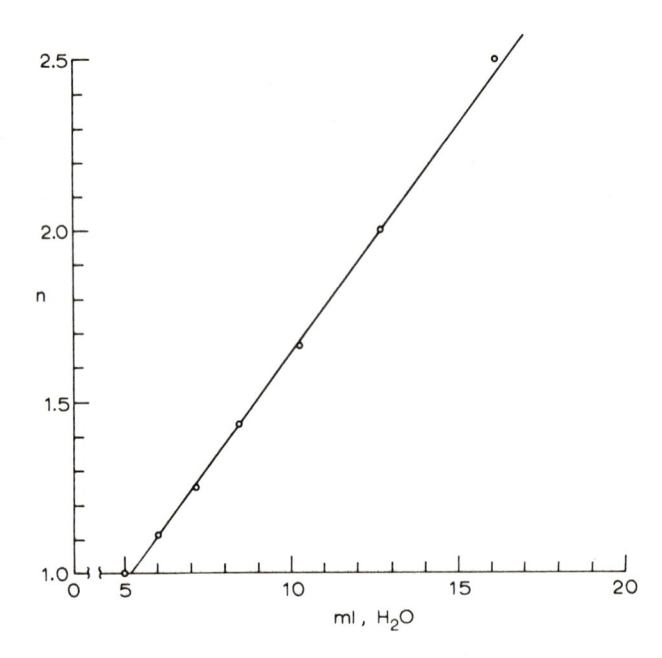

FIG. 15. PLOT OF THE DEGREE OF DILUTION, n, VERSUS THE VOLUME OF TITRANT FOR THE SYSTEM BENZENE-ABSOLUTE ETHANOL TITRATED WITH WATER.

is chosen which is totally miscible with A but only partially miscible with B. Enough C is added to all samples such that the composition of the ternary mixtures lie along a line of constant per cent C which passes close to the plait point (line B'T in Fig. 16). The system will have two phases for all mixtures of A and B between compositions, pure B and S_M. The two-phase system is titrated with a mixture of C and A of composition T to the disappearance of turbidity. Line B'T is therefore the titration line. It passes through the binodal

2. UNUSUAL END-POINT DETECTION METHODS

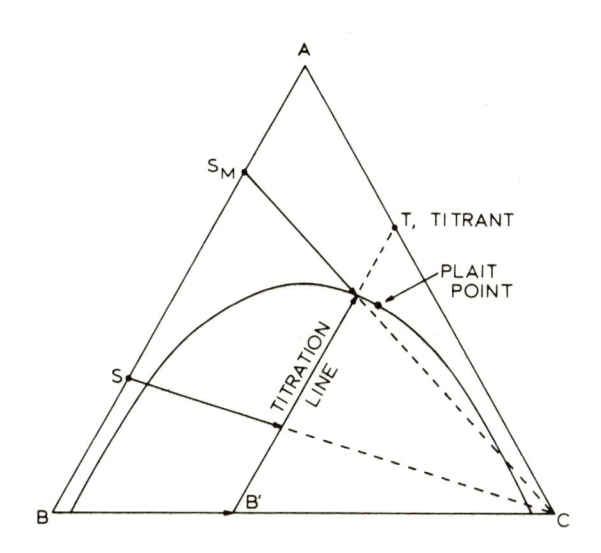

FIG. 16. TERNARY LIQUID-LIQUID MISCIBILITY DIAGRAM ILLUSTRATING THE ANALYSIS OF THE BINARY SYSTEM A-B BY PHASE TITRATION WITH CLARIFICATION END POINT. ADAPTED FROM REF. [41] WITH PERMISSION.

curve at the same point (close to the plait point) for all samples. In this manner, the best possible precision can be achieved and it is independent of the composition of the original sample. The following equation for calculating the volume fraction F_A of component A in the sample is given by Dunnery and Atwood [41]:

$$F_A = (V_{tB} - V_{tS})/(V_{tB} + V_S)$$

where V_S is the volume of sample taken plus the volume of C added, V_{tB} is the volume of titrant needed for a sample of pure B, and V_{tS} is the volume of titrant needed for the sample. For a given experiment, V_{tB} and V_S are constants and Eq. (51) describes a linear relationship between the volume of titrant required for the sample and F_A. Figure 17 shows a comparison of the linear and the Type A calibration curves for the system of Fig. 16. It is seen that the range of sample compositions that can be handled in a simple manner by the linear technique, is considerably larger than that for the turbidity end-point method. At F_A = 0 and 1, the corresponding volume axis coordinates are $V_{tS} = V_{tB}$ and $-V_S$ so in principle only one experimental data point is required to establish the calibration curve. Dunnery and Atwood [41] point out that secondary effects such as volume changes due to nonideal mixing,

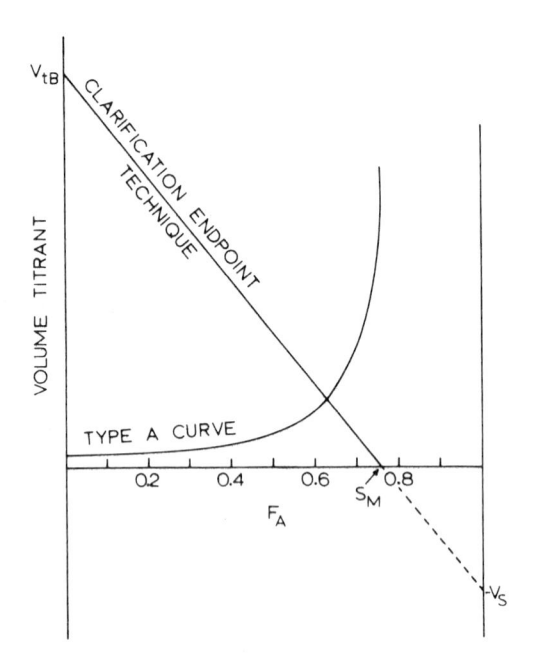

FIG. 17. A COMPARISON OF THE THEORETICAL CALIBRATION CURVES FOR THE TURBIDIMETRIC AND CLARIFICATION END-POINT PHASE TITRATION TECHNIQUES. ADAPTED FROM REF. [41] WITH PERMISSION.

2. UNUSUAL END-POINT DETECTION METHODS

temperature effects, errors in preparing the titrant, and traces of impurities in the unknown or titrant can cause deviations from Eq. (51). These are usually small and, within the precision of the measurements can be considered linear. Their net effect is to change the slope of the calibration line and a number of points should be obtained.

Yet another approach for the analysis of binary mixtures is due to Rogers et al. [42]. The objective is to titrate mixtures of chemically similar substances which even have nearly the same solubility in water. The procedure involves adding a constant amount of a third component to the binary solution and titrating the resulting ternary solution with water to a turbidimetric end point. An understanding of the process can be gained from Figs. 18 and 19. The system under

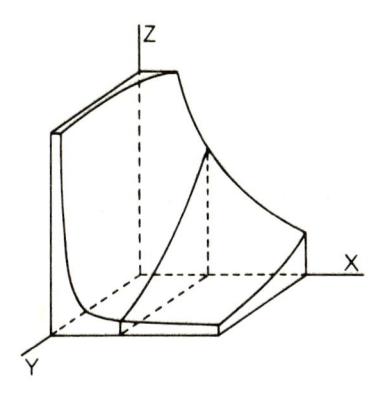

FIG. 18. THE SOLUBILITY SURFACE FOR THE TERNARY SYSTEM CARBON TETRACHLORIDE-CHLOROFORM-ETHANOL TITRATED WITH WATER. THE AXES ARE: X, 100% ETHANOL; Y (MILLILITERS OF CARBON TETRACHLORIDE/MILLILITERS OF CARBON TETRACHLORIDE + MILLILITERS OF CHLOROFORM), 100%, i.e., CARBON TETRACHLORIDE BEFORE THE ADDITION OF ANY ETHANOL OR WATER; AND Z, MILLILITERS OF WATER NECESSARY TO CAUSE TURBIDITY. FROM REF. [42] WITH PERMISSION.

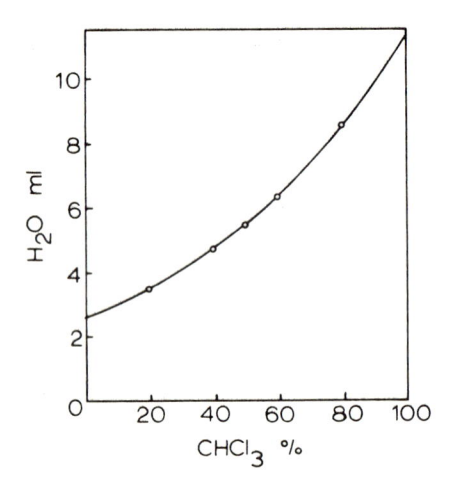

FIG. 19. TYPE B CALIBRATION CURVE FOR THE SYSTEM CARBON TETRACHLORIDE-CHLOROFORM. THE HORIZONTAL AXIS REPRESENTS THE PER CENT CHLOROFORM BEFORE ADDITION OF ETHANOL. FROM REF. [42] WITH PERMISSION.

consideration is a binary mixture of chloroform and carbon tetrachloride with added ethanol. The three-dimensional surface shown in Fig. 18 is the solubility surface for the milliliters of water necessary to titrate the three-component system, CCl_4, $CHCl_3$, EtOH. The plane shown parallel to the YZ axis represents ternary solutions of compositions with a constant per cent ethanol. The curve defined by the intersection of this plane with the solubility surface of Fig. 18, when projected on the YZ plane, yields the calibration curve shown in Fig. 19. Defining the Y axis of Fig. 18 in terms of per cent CCl_4 in the original binary solution automatically defines the horizontal axis of the calibration curve in terms of either per cent CCl_4 or per cent $CHCl_3$. Rogers and co-workers refer

2. UNUSUAL END-POINT DETECTION METHODS

to this calibration curve as a Type B calibration curve. It is noted that the projections on the XZ plane of the inner and outer edges of the solubility surface parallel to the XZ plane give the Type A calibration curves for the systems $CHCl_3$-EtOH and CCl_4-EtOH, respectively, when titrated with water. The shape of the Type B calibration curve depends on the choice of the amount of misicible component added. Rogers et al. [42] suggest that it be chosen to produce limits for the water titer of about 1 and 12 ml. Details on this point can be found in the original article.

3. *Methods for Ternary Solutions*

Bogin very early described a procedure for the analysis of ternary mixtures [24]. One of the components is determined by an independent method of analysis. A series of standard solutions are then prepared containing this same amount of the component found by analysis and varying ratios of the remaining two components. The standards are then titrated with water to a turbidity end point and a working curve prepared from the data. To save time, a family of such curves is prepared; each curve corresponding to a fixed percentage of the component determined by independent analysis. An analysis of an unknown proceeds by first running the independent analysis to establish which curve of the family of curves is to be used. Then, the titration is performed on a fresh portion of sample to which the appropriate amount of the third component has been added. The percentage of the two remaining components is then determined from the calibration curve. This method is very much related to the technique for binary solutions of Rogers and co-workers involving Type B calibration curves. In fact, Bogin's calibration curves are Type B and it follows that a series of such curves is needed to extend the method for binary mixtures to one for ternary mixtures.

Siggia and Hanna [43] developed a method for ternary solutions which eliminated the calibration curves but still required an independent analysis for one of the components. To do this, a number of weighings are necessary and the phase diagram for the system must be accurately known. The independent analysis is first carried out and expressed in terms of weight per cent of, let us say, component A. A weighed amount of the sample is then titrated to a turbidimetric end point and the weight of titrant (component B) delivered is calculated from a weighing of the end-point solution. The per cent of A in the solution at the end point (%A_{ep}) is calculated as follows:

%A_{ep} = 100 (%A in original sample)(wt of sample) / wt of ep solution   (52)

The composition of the titrated solution at the end point corresponds for all practical purposes to a point on the binodal curve. The %A_{ep} is the A coordinate of this point and the remaining two coordinates can be obtained from the phase diagram, i.e., %B_{ep} and %C_{ep}. With this information, the per cent of B and the per cent of C in the original sample are calculated:

%B = 100 (%B_{ep})(wt of ep solution)-wt of B delivered / sample wt   (53)

%C = 100 (%C_{ep})(wt of ep solution) /sample wt   (54)

The phase diagram itself is prepared by titrating known mixtures of the two miscible components of the system with the third component until a turbidimetric end point is reached. The composition of the solution at the end point is calculated and the point plotted on the diagram.

Our discussion of methods will be completed by presenting the technique of Suri [44] which is of interest because it

2. UNUSUAL END-POINT DETECTION METHODS 151

eliminates the independent analysis for one of the components. Let A and B be the immiscible pair of the ternary system A, B, and C, as shown in Fig. 20. Consider the point, P, located on the left-hand side of the binodal curve. If C is now added to the solution, the composition will now be given by a point, Q, on the line joining the point P with the C apex, and the solution will be homogeneous. This solution is now titrated with B to a turbidity end point, which will correspond to a point lying on the binodal curve and on the line joining the B apex with Q. Let the weight of sample taken of composition P be W_s = 1.000 g, and the weight of exactly 1 ml of C at the temperature of the experiment be W_c. The weight of the solution at the end point is W_{ep}, the weight of B added is W_b and the weight of A in the original sample is W_a. Then, the weight

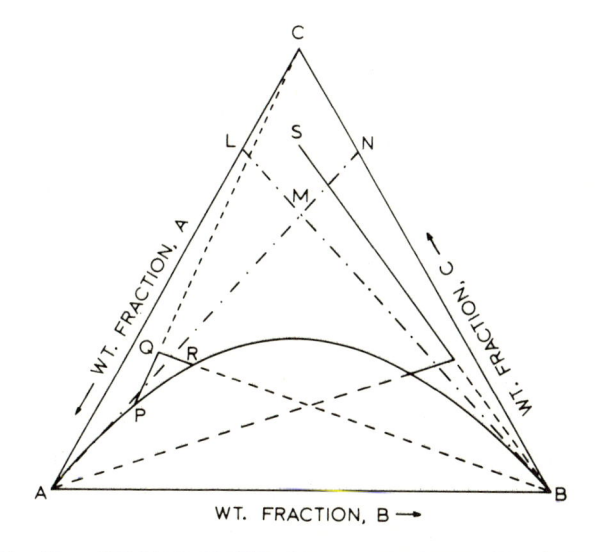

FIG. 20. TERNARY LIQUID-LIQUID MISCIBILITY DIAGRAM ILLUSTRATING THE ANALYSIS OF THE TERNARY MIXTURE, S, BY PHASE TITRATIONS. ADAPTED FROM REF. [44] BY PERMISSION.

fraction of A at point P, $X_{A,P}$, is given by
$$X_{A,P} = W_a/W_s \tag{55}$$
and the weight fraction of A at the end point, $X_{A,ep}$, by
$$X_{A,ep} = W_a/W_{ep} \tag{56}$$
Then,
$$X_{A,P}/X_{A,ep} = W_{ep}/W_s \tag{57}$$
But
$$W_s = 1.000 \text{ g and } W_{ep} = 1 + W_b + W_c \tag{58}$$
Substituting Eq. (58) in Eq. (57) and rearranging
$$W_b = X_{A,P}/X_{A,ep} - (1 + W_c) \tag{59}$$
Since there are two phases present at point P and temperature and pressure are constant, the system is invariant when the weight fraction of A at P is fixed. With W_s and W_c constant, a calibration curve of W_b versus weight fraction of A can be prepared. Similarly, a second calibration curve of W_a versus weight fraction of B is obtained. A series of points P along the binodal curve are chosen and the composition of the mixture for each of these points is read from the phase diagram. A known weight of C is then chosen and the points Q corresponding to points P are calculated. The lines QA or QB are then drawn on the phase diagram and the points R (see Fig. 20), corresponding to the end-point solutions, are obtained. The weight fractions $X_{A,P}$ or $X_{B,P}$ and $X_{A,ep}$ or $X_{B,ep}$ are then calculated with the information obtained from the phase diagram. Equation (59) is then used to calculate W_b or W_a and the calibration curves are prepared. The analysis of any unknown solution whose composition happens to lie on the binodal curve is straightforward. The known weight of C is added and W_b or W_a found as described above using the calibration curves. With either one, point P has now been located and the phase diagram is used to find the weight fractions of the remaining two components.

2. UNUSUAL END-POINT DETECTION METHODS 153

If the solution is homogeneous, the approach is to change the composition of the original solution so that a new solution is produced which will have a composition corresponding to a point P on the binodal curve. Once this is done, the new solution serves as a starting point for the procedure given above. From the composition found for point P, the composition of the original solution is obtained by calculation. Two different cases for homogeneous unknowns can be distinguished. If the solution composition corresponds to a point with area CLMN of Fig. 20, it is only possible to reach a point, P, on the binodal curve by appropriate addition of both A and B. This is illustrated for point S in the figure. A known weight of B is added to a known weight of S to bring the composition into area BMN. The weight of A then required to produce turbidity is measured and the solution composition has been brought to a point P on the binodal curve. The grams of A present for composition S is equal to the grams of A present at P less the weight of A added. A similar procedure is followed to find the grams of B present at S, and C is calculated by difference. In the second case, the unknown solution composition lies in the area bounded by the binodal curve and ALMNB of Fig. 20. In this region, only A or B is added to bring the composition onto the binodal curve.

C. Apparatus and Experimental Methods

The basic equipment requirements for phase titrations are a beaker or flask to hold the sample and a buret. Considering that water might be the titrant and that a chemical indicator may not be required, it is difficult to imagine a more economical method of chemical analysis! However, several other factors enter the picture which may raise the cost slightly.

Stirring is universally recommended. Caley and Habboush [30] suggest slow stirring to avoid producing bubbles which obscure the end point. On the other hand, Rogers [27] recommends violent stirring. The difference is probably that the former authors used iodine as a chemical indicator and bubble formation was a serious difficulty to them. The solution temperature must be considered and the most serious sources of temperature problems are room temperature changes and heat effects due to the addition of titrant. The problem of temperature control has been handled in several ways. Bogin [24] prepared a series of calibration curves at different temperatures and then conducted his titrations without temperature control. The temperature of the solution was measured at the end point and the composition of the unknown determined from the appropriate working curve or by interpolation between curves if necessary. Other workers have used constant temperature baths in the preparation of the working curves and in the actual titrations. Caley and Habboush [30] used a water-jacketed buret through which water from the bath was circulated. Theirs was a 50-ml buret. Others used without water jackets include 2-ml microburets and 0.1-ml ultramicroburets. Since calibration curves are easily and quickly prepared, Rogers [27] adopted the practical approach of neither measuring the temperature nor attempting to control it other than by working in a laboratory with some degree of room temperature regulation. Data for calibration curves and for titrations were obtained under as identical conditions as possible and if some change in conditions were suspected, a new calibration curve was determined. Finally, good illumination is of great help in identifying the end point. Rogers recommends cross illumination and a black background [27]. Caley and Habboush used a circular fluorescent lamp [30].

2. UNUSUAL END-POINT DETECTION METHODS   155

Rogers et al. [45] have reported an apparatus for titrating very small volumes of sample. Small test tubes or centrifuge tubes were used as sample holders. Stirring was done magnetically or by a vibrating wire. The sample holder was placed in a black box and illuminated from the side through the means of a Lucite rod to transmit light from the source. The latter feature avoided heating the sample. The sample was viewed through a blue light filter mounted on the front of the box. Sample volumes less than 0.5 ml were titrated with ultramicroburets.

D. Applications

Table 5 summarizes the systems to which phase titrations have been applied. The table is divided into three sections: binary solutions of nonelectrolytes, ternary solutions of nonelectrolytes, and solutions of electrolytes. The column headed "best range" has several meanings. For Type A calibration curves, it refers to the optimum range suggested by the authors of the work or, if this information was not provided, to the best guess of this writer. For linear working curves, it refers to the range over which the curve is linear. In some cases, the information was not known at all or not applicable. Brief comments in column 5 attempt to clarify each situation. The table undoubtedly does not list every phase titration reported in the literature but the attempt has been made to make it as complete as possible. The possible scope of phase titrations is indicated by the wide variety of samples found in Table 5. Examples include mixtures of interest to the petroleum, cosmetic, and solvents industries. The most important applications appear to be those where mixtures of very similar liquids are determined. In general, it is

TABLE 5

Applications of Phase Titrations

Number	Sample	Titrant	Best range (%)a	Comments	Reference
A. Binary mixtures of nonelectrolytes
1	n-Butyl alcohol-ethanol	Water	25-50	Type A curve	24
2	n-Butyl alcohol-n-propanol	Water	35-55	Type A curve	24
3	n-Butyl alcohol-isopropanol	Water	45-65	Type A curve	24
4	Benzene-ethanol	Water	50-90	Type A curve	24
5	Petroleum ether-ethanol	Water	70-90	Type A curve	24
6	Gasoline-ethanol	Water	75-93	Type A curve	24
7	Kerosene-ethanol	Water	86-94	Type A curve	24
8	Methanol-benzene	Water	5-40	Type A curve	30
9	Methanol-toluene	Water	5-40	Type A curve	30
10	Methanol-xylene	Water	5-35	Type A curve	30
11	Methanol-mesitylene	Water	5-30	Type A curve	30
12	Ethanol-benzene	Water	5-45	Type A curve	30
13	Ethanol-toluene	Water	5-45	Type A curve	30

2. UNUSUAL END-POINT DETECTION METHODS

14	Ethanol-xylene	Water	5-40	Type A curve	30
15	Ethanol-mesitylene	Water	5-40	Type A curve	30
16	Ethanol-p-cymene	Water	5-40	Type A curve	30
17	isopropanol-benzene	Water	10-40	Type A curve	30
18	isopropanol-toluene	Water	10-40	Type A curve	30
19	isopropanol-xylene	Water	10-40	Type A curve	30
20	isopropanol-mesitylene	Water	10-40	Type A curve	30
21	isopropanol-p-cymene	Water	10-40	Type A curve	30
22	CCl_4-methanol	Water	2-30	Type A curve	27
23	CCl_4-ethanol	Water	3-50	Type A curve	27
24	CCl_4-isopropanol	Water	10-50	Type A curve	27
25	CCl_4-dioxane	Water	2-20	Type A curve	27
26	CCl_4-acetone	Water	3-25	Type A curve	27
27	CS_2-methanol	Water	2-20	Type A curve	27
28	CS_2-ethanol	Water	3-20	Type A curve	27
29	CS_2-isopropanol	Water	20-40	Type A curve	27
30	CS_2-dioxane	Water	3-20	Type A curve	27
31	CS_2-acetone	Water	3-25	Type A curve	27
32	Nitroethane-methanol	Water	25-70	Fair-poor end points	46
33	Nitroethane-ethanol	Water	30-60	Fair-poor end points	46
34	Nitroethane-1-propanol	Water	30-60	Fair-poor end points	46

Table 5--Continued

Number	Sample	Titrant	Best range (%)[a]	Comments	Reference
35	Nitroethane-2-propanol	Water	30-60	Fair-poor end points	46
36	Nitroethane-acetone	Water	18-35	Fair-poor end points	46
37	Nitroethane-dioxane	Water	20-40	Fair-poor end points	46
38	Nitroethane-acetic acid	Water	35-70	Fair-poor end points	46
39	Nitromethane-methanol	Water	--	Very poor end point	46
40	1-Nitropropane-methanol	Water	14-55	Type A, fair end point	46
41	1-Nitropropane-ethanol	Water	15-60	Type A, fair end point	46
42	1-Nitropropane-1-propanol	Water	20-50	Type A, fair end point	46
43	1-Nitropropane-2-propanol	Water	20-50	Type A, fair end point	46
44	1-Nitropropane-acetone	Water	10-30	Type A, fair end point	46
45	1-Nitropropane-dioxane	Water	12-30	Type A, fair end point	46
46	2-Nitropropane-methanol	Water	13-65	Type A, fair end point	46
47	2-Nitropropane-ethanol	Water	14-63	Type A, fair end point	46
48	2-Nitropropane-1-propanol	Water	20-65	Type A, fair end point	46
49	2-Nitropropane-2-propanol	Water	24-58	Type A, fair end point	46
50	2-Nitropropane-acetone	Water	10-35	Type A, fair end point	46
51	2-Nitropropane-dioxane	Water	11-35	Type A, fair end point	46

2. UNUSUAL END-POINT DETECTION METHODS

52	Pyridine-benzene	Water	12-45	Type A curve	47
53	Pyridine-toluene	Water	7-45	Type A curve	47
54	Pyridine-xylene	Water	5-45	Type A curve	47
55	Pyridine-nitromethane	Water	45-70	Type A curve	47
56	Pyridine-2-nitropropane	Water	30-65	Type A curve	47
57	Pyridine-cyclohexane	Water	2-20	Type A curve	47
58	Pyridine-bromobenzene	Water	8-45	Type A curve	47
59	Pyridine-CCl$_4$	Water	3-35	Type A curve	47
60	Pyridine-chloroform	Water	12-40	Type A curve	47
61	Pyridine-1-chlorobutane	Water	10-40	Type A curve	47
62	Pyridine-bromoethane	Water	15-40	Type A curve	47
63	Ethanol-methyl salicylate	Water	6-35	Type A curve	45
64	Ethanol-aldehyde C-8	Water	8-58	Type A curve	45
65	Ethanol-aldehyde C-12	Water	10-30	Type A curve	45
66	Ethanol-Geraniol Coeur	Water	8-55	Type A curve	45
67	Ethanol-phenylethylether	Water	38-83	Type A curve	45
68	Ethanol-phenylethylacetate	Water	5-50	Type A curve	45
69	Ethanol-geranylacetate	Water	9-50	Type A curve	45
70	Ethanol-lemongrass oil	Water	2-45	Type A curve	45
71	Ethanol-citronella oil	Water	1-55	Type A curve	45
72	Ethanol-α-ionone	Water	4-55	Type A curve	45

Table 5--Continued

Number	Sample	Titrant	Best range (%)a	Comments	Reference
73	Ethanol-β-ionone	Water	1-50	Type A curve	45
74	Ethanol-water	Water	--	Furfural indicator	48
75	Acetone-water	Water	--	Furfural indicator	48
76	Acetone-methanol	Water	0-10	Furfural indicator	31
77	Acetone-methylacetate	Water	0-10	Furfural indicator	31
78	Methanol-methylacetate	Water	0-35	Furfural indicator	31
79	Ethanol-isoamylalcohol	Water	--	Furfural indicator	32
80	Ethanol-acetone	Water	--	Furfural indicator	32
81	Acetone-methylacetate	Water	--	Furfural indicator	32
82	Ethanol-methanol	Water	--	Furfural indicator	33
83	Ethanol-n-butanol	Water	--	Furfural indicator	33
84	Ethanol-water	Water	--	Organic indicators	35
85	Methanol-water	Water	--	Organic indicators	35
86	Acetone-water	Water	--	Organic indicators	35
87	Acetone-methanol	Water	--	Organic indicators	35
88	Acetone-methylacetate	Water	--	Organic indicators	35
89	Ethanol-isoamyl alcohol	Water	15-40	Linearized curve	40

2. UNUSUAL END-POINT DETECTION METHODS

90	Dioxane-isoamyl alcohol	Water	25-50	Linearized curve	40
91	Ethanol-acetoacetic ester	Water	5-10	Linearized curve	49
92	Isopropanol-acetoacetic ester	Water	5-10	Linearized curve	49
93	Acetone-acetoacetic ester	Water	--	Linearized curve	49
94	Dioxane-acetoacetic ester	Water	--	Linearized curve	49
95	Aniline-methanol	Water	40-80	Type A curve	28
96	Aniline-ethanol	Water	50-85	Type A curve	28
97	Aniline-isopropanol	Water	40-80	Type A curve	28
98	Aniline-dioxane	Water	30-50	Type A curve	28
99	Aniline-acetone	Water	20-50	Type A curve	28
100	Nitrobenzene-methanol	Water	4-25	Type A curve	28
101	Nitrobenzene-ethanol	Water	5-30	Type A curve	28
102	Nitrobenzene-isopropanol	Water	10-40	Type A curve	28
103	Nitrobenzene-dioxane	Water	5-25	Type A curve	28
104	Nitrobenzene-acetone	Water	5-25	Type A curve	28
105	Methyl aniline-methanol	Water	8-30	Type A curve	28
106	Methyl aniline-acetone	Water	8-30	Type A curve	28
107	Methyl aniline-isopropanol	Water	15-45	Type A curve	28
108	Turpentine-acetone	Water	2-15	Type A curve	28
109	Ethylacetate-acetone	Water	30-55	Type A curve	28
110	n-Butyl bromide-methanol	Water	2-20	Type A curve	28

Table 5--Continued

Number	Sample	Titrant	Best range (%)a	Comments	Reference
111	n-Butyl bromide-ethanol	Water	3-30	Type A curve	28
112	n-Butyl bromide-isopropanol	Water	10-40	Type A curve	28
113	n-Butyl bromide-dioxane	Water	2-20	Type A curve	28
114	n-Butyl bromide-acetone	Water	3-20	Type A curve	28
115	Dichloromethane-methanol	Water	13-50	Type A curve	42
116	Dichloromethane-ethanol	Water	15-50	Type A curve	42
117	Dichloromethane-isopropanol	Water	17-38	Type A curve	42
118	Dichloromethane-dioxane	Water	7-22	Type A curve	42
119	Dichloromethane-acetone	Water	6-20	Type A curve	42
120	Dichloromethane-acetic acid	Water	15-50	Type A curve	42
121	Cyclohexane-methanol	Water	1-10	Type A curve	42
122	Cyclohexane-ethanol	Water	1-20	Type A curve	42
123	Cyclohexane-isopropanol	Water	5-45	Type A curve	42
124	Cyclohexane-dioxane	Water	1-15	Type A curve	42
125	Cyclohexane-acetone	Water	1-25	Type A curve	42
126	Cyclohexane-acetic acid	Water	1-7	Type A curve	42
127	Allyl bromide-methanol	Water	4-35	Type A curve	42

2. UNUSUAL END-POINT DETECTION METHODS

128	Allyl bromide-ethanol	Water	5-40	Type A curve	42
129	Allyl bromide-isopropanol	Water	10-45	Type A curve	42
130	Allyl bromide-dioxane	Water	4-25	Type A curve	42
131	Allyl bromide-acetone	Water	3-20	Type A curve	42
132	Allyl bromide-acetic acid	Water	5-30	Type A curve	42
133	Benzene-dioxane	Water	3-20	Type A curve	42
134	Benzene-acetone	Water	4-25	Type A curve	42
135	Benzene-acetic acid	Water	5-30	Type A curve	42
136	Toluene-dioxane	Water	2-20	Type A curve	42
137	Toluene-acetone	Water	3-23	Type A curve	42
138	Toluene-acetic acid	Water	2-25	Type A curve	42
139	Xylene-dioxane	Water	1-20	Type A curve	42
140	Xylene-acetone	Water	2-25	Type A curve	42
141	Xylene-acetic acid	Water	2-20	Type A curve	42
142	Bromobenzene-methanol	Water	2-25	Type A curve	50
143	Bromobenzene-ethanol	Water	3-33	Type A curve	50
144	Bromobenzene-isopropanol	Water	7-45	Type A curve	50
145	Bromobenzene-dioxane	Water	2-20	Type A curve	50
146	Bromobenzene-acetone	Water	2-25	Type A curve	50
147	Bromobenzene-acetic acid	Water	2-23	Type A curve	50
148	Chloroform-methanol	Water	8-60	Type A curve	50
149	Chloroform-ethanol	Water	11-60	Type A curve	50

Table 5--Continued

Number	Sample	Titrant	Best range (%)[a]	Comments	Reference
150	Chloroform-isopropanol	Water	12-40	Type A curve	50
151	Chloroform-dioxane	Water	4-20	Type A curve	50
152	Chloroform-acetone	Water	4-30	Type A curve	50
153	Chloroform-acetic acid	Water	10-50	Type A curve	50
154	1,2-Dibromoethane-methanol	Water	2-25	Type A curve	50
155	1,2-Dibromoethane-ethanol	Water	2-20	Type A curve	50
156	1,2-Dibromoethane-isopropanol	Water	5-25	Type A curve	50
157	1,2-Dibromoethane-dioxane	Water	2-15	Type A curve	50
158	1,2-Dibromoethane-acetone	Water	3-20	Type A curve	50
159	1,2-Dibromoethane-acetic acid	Water	3-25	Type A curve	50
160	Ethanol-benzene	Water	40-60	Type A curve	28
161	Methylaniline-aniline, added ethanol	Water	--	Type B curve	42
162	CCl$_4$-CHCl$_3$, added acetic acid	Water	--	Type B curve	42
163	n-Butyl bromide-n-butyl acetate, added ethanol	Water	--	Type B curve	42
164	Nitrobenzene-aniline, added ethanol	Water	--	Type B curve	42

2. UNUSUAL END-POINT DETECTION METHODS

165	Cyclohexane-benzene, added ethanol	Water	--	Type B curve	42
166	Cyclohexane-benzene, added: 90% ethanol, 10% water	Water	--	Type B curve	42
167	Methanol-n-propanol, added cyclohexane	Water	--	Type B curve	47
168	Ethanol-n-propanol, added cyclohexane	Water	--	Type B curve	47
169	Methanol-ethanol, added cyclohexane	Water	--	Type B curve	47
170	n-Butanol-ethanol	5% aq. NaCl solution	40-60	Type A curve	24
171	n-Butanol-ethanol	20% aq. NaCl solution	70-85	Type A curve	24
172	n-Butyl acetate-n-butanol	60% n-BuOH 15% EtOH 25% H$_2$O	30-50	Type A curve	24
173	n-Butyl acetate-n-butanol	85% n-BuOH 15% H$_2$O	30-50	Type A curve	24
174	Benzene-ethanol	50% EtOH 50% H$_2$O	30-60	Type A curve	24
175	Acetone-isobutyl methyl ketone	67% acetone 33% H$_2$O	--	Clarification e.p.	41
176	Ethanol-water	n-BuOH	--	Type A curve	24

Table 5--Continued

Number	Sample	Titrant	Best range (%)[a]	Comments	Reference
177	n-Butanol-water	Benzene	3-9	Type A curve	24
178	Acetone-water	Benzene	3-9	Type A curve	24
179	Acetone-water	CCl_4	4-20	Type A curve	27
180	CCl_4-water	Acetone	3-10	Type A curve	27
181	Dioxane-water	Benzene	20-30	Type A curve	28
182	Pyridine-water	$CHCl_3$	8-45	Type A curve	50
B. Ternary mixtures of nonelectrolytes
183	Ethanol-acetone-butanol	Water	22-36	Acetone determined independently	24
184	Benzene-cyclohexane-nitromethane	Cyclohexane or nitromethane	--	Cyclohexane is the best titrate	44
185	Benzaldehyde-dioxane-water	Water	--	Benzaldehyde determined independently	43
186	Benzaldehyde-dioxane-water	Benzaldehyde	--	Benzaldehyde determined independently	43
187	Water-pyridine-benzene	Water	--	Water determined independently	43

2. UNUSUAL END-POINT DETECTION METHODS

188	Water-pyridine-benzene	Benzene	--	Water determined independently	43
189	Ethyl vinyl ether-water-ethanol	Ethanol	--	Ethyl vinyl ether found independently	43
190	Methanol-chlorobenzene-water	Water	--	Water determined independently	43
191	Acetic acid-CCl$_4$-water	Water	--	Acetic acid found independently	43
192	Glycerol-acetic acid-benzene	Benzene	--	Glycerol determined independently	43
193	Monoethanolamine-pyridine-ethyl ether	Ethyl ether	--	Monoethanolamine found independently	43

C. Solutions of electrolytes

194	NH$_4$Cl in water	Acetone	8-18	Type A curve	24
195	Na$_2$SO$_4$ in water	Ethanol	--	--	51
196	MgSO$_4$ in water	Ethanol	--	--	51
197	Na$_2$CO$_3$ in water	Ethanol	--	--	51
198	Na$_2$S$_2$O$_3$ in water	Ethanol	--	--	51
199	NH$_4$Cl in water	Water	--	Isobutanol indicator	34
200	MgCl$_2$ in water	Water	--	Isobutanol indicator	34
201	FeCl$_3$ in water	Water	--	Isobutanol indicator	34
202	KCl in water	Water	--	Isobutanol indicator	34

Table 5--Continued

Number	Sample	Titrant	Best range (%)[a]	Comments	Reference
203	CoCl$_2$ in water	Water	--	Isobutanol indicator	34
204	BaCl$_2$ in water	Water	--	Isobutanol indicator	34
205	CaCl$_2$ in water	Water	--	Isobutanol indicator	34
206	NaCl in water	Water	--	Isobutanol indicator	34
207	NiCl$_2$ in water	Water	--	Isobutanol indicator	34

[a] Per cent of second compound listed under sample except for No. 194 where the per cent is that of NH$_4$Cl.

2. UNUSUAL END-POINT DETECTION METHODS

interesting that other methods of titration are applicable to mixtures only in special cases and/or with difficulty, but phase titrations are uniquely based on the fact that the sample taken for analysis must be a mixture.

Phase titrations are a convenient way to determine the water in organic solvents. Excellent results were obtained by Rogers and Ozsogomonyan [50] for the determination of water in pyridine by titration with chloroform. Samples containing from 10 to 40% water were titrated with an average error of 0.07% absolute and a maximum error of 0.2% absolute. Other solvents examined and the titrants used are listed in Table 5 (Nos. 176-181). Unfortunately, examination of the best range column of Table 5 suggests that the method is applicable only when the volume per cent of water is 3% or greater. It would be valuable if the method could be extended to the determination of millimolar or smaller concentrations of water in organic solvents.

Another area of application is that of electrolyte solutions. The solubility relationships of aqueous solutions of electrolytes in the presence of organic solvents are too diverse to present in the manner used for nonelectrolytes in the methods section. However, the analysis of the salt concentration of an aqueous solution is possible by phase titration. Bogin seems to have been the first to do this [24]. The system studied was ammonium chloride in water titrated with acetone. A calibration curve resembling a Type A curve was found with the best range being about 8 to 18% ammonium chloride. Spiridonova has done extensive work along these lines [37,51-54] and some of the systems he has studied are listed in Table 4 (Nos. 195-207). In a further extension of this idea, Spiridonova has determined the moisture content of crystalline hydrates by titration with water in the presence

TABLE 6

Analysis of the System
Methyl Alcohol-Chlorobenzene-Water[a]

Sample no.	% Methyl alcohol Found	Present	% Chlorobenzene Found	Present	% Water Found	Present
1	70.0	70.17	4.60	4.97	25.68	24.86
2	78.8	80.07	10.7	9.97	10.30	9.96
3	60.5	59.08	34.0	35.90	5.6	5.03
4	52.3	50.99	41.9	44.00	5.7	5.01
5	31.9	30.73	65.9	67.31	2.3	1.96
6	83.2	82.01	1.67	3.00	15.2	14.95
7	75.3	75.46	19.5	19.49	5.2	5.06

[a]From Ref. [43], with permission.

of ethanol-isobutanol mixture [55]. Errors ranged from 0.3 to 2.8%.

The accuracy and precision of a phase titration depend not only on the usual factors affecting any volumetric procedure but also on the sharpness of the end point and the composition of the sample. These last two factors are interrelated and are the reason for the using the concept of the optimum range to describe the applicability of a phase titration to a given system. With the best conditions prevailing, accuracies and precisions of a few parts per thousand are entirely possible. The data of Siggia and Hanna [43] for the ternary system methanol-chlorobenzene-water are shown in Table 6. Water was determined independently by the Karl Fischer method and was used as the titrant. Quite a variation in precision and accuracy for methanol and chlorobenzene is evident but the

2. UNUSUAL END-POINT DETECTION METHODS 171

TABLE 7

Titration of Binary Systems with Water[a]

Sample	Optimum range (%)[b]	Number of titrations	Average error absolute %
Aniline-methanol	40-80	12	0.42
Aniline-isopropanol	40-80	12	0.18
Aniline-acetone	20-50	12	0.38
Nitrobenzene-dioxane	5-25	12	0.09
Nitrobenzene-acetone	5-25	18	0.20
Methyl aniline-acetone	8-30	14	0.12
n-Butyl bromide-methanol	2-20	12	0.11

[a] Extracted from Ref. [28], with permission.
[b] Refers to second compound listed under each sample.

results for sample 7 are correct within one or two parts per thousand for each of these compounds. In view of the variations in relative accuracy for a given sample, Rogers and Ozsogomonyan [28] have expressed accuracy as the average error in absolute per cent over the optimum range. Data extracted from their paper are shown in Table 7. Again it is seen that relative accuracy varies considerably but excellent results were obtained in some cases. Roger's approach of not controlling temperature but running frequent calibration curves is successful in view of the data in Table 7. Further, a relative accuracy of several per cent may be entirely adequate for the analysis and much of the data of Tables 6 and 7 fall in this range. For the clarification end-point method of Caley and Habboush [30] where a linear calibration curve is found, the absolute standard deviation for all samples of the acetone-isobutylmethylketone system studied was 0.07 ml of titrant (33.3% v/v water and acetone). As little as several milliliters of titrant were required for some sample compositions, but for many compositions, much larger volumes were required and the precision is excellent.

In summary, phase titrations offer a useful method of analysis for a wide variety of samples. Only a fraction of the number of binary and ternary solutions which could possibly be analyzed by this method have been examined, and the technique could probably be applied to the assay of pure liquids by deliberate addition of a second solvent.

V. FLAME PHOTOMETRIC TITRATIONS

A. Introduction

Since flame photometry is generally regarded as useful in the parts per million concentration range, it might be expected that the method would be widely used as an end-point detection technique for titrimetry. The process would involve following the emission intensity of some flame species, which corresponds to one of the reactants or products of a titration reaction, as a function of the volume of titrant added. The resulting titration curve would probably be of the linear-segmented type, at least for data corresponding to points not too far removed from the end point. A difficulty is that the measured emission intensity must be related to only one of the species taking part in the titration reaction, unless the unlikely circumstance existed where two species in solution gave rise to the same emission line but the intensity of emission, as a function of solution concentration, was quite different for

2. UNUSUAL END-POINT DETECTION METHODS

each species. This problem no doubt is responsible for the lack of a significant number of papers in the literature.

Thus far, the only published flame photometric titration procedures involve radiation interferences. The depression of the calcium emission intensity by phosphate is well known. Figure 21, taken from the work of Yofe and Finkelstein [56], illustrates the phenomenon and is typical of other results reported in the literature. It shows that the calcium emission intensity decreases to a value which is independent of the

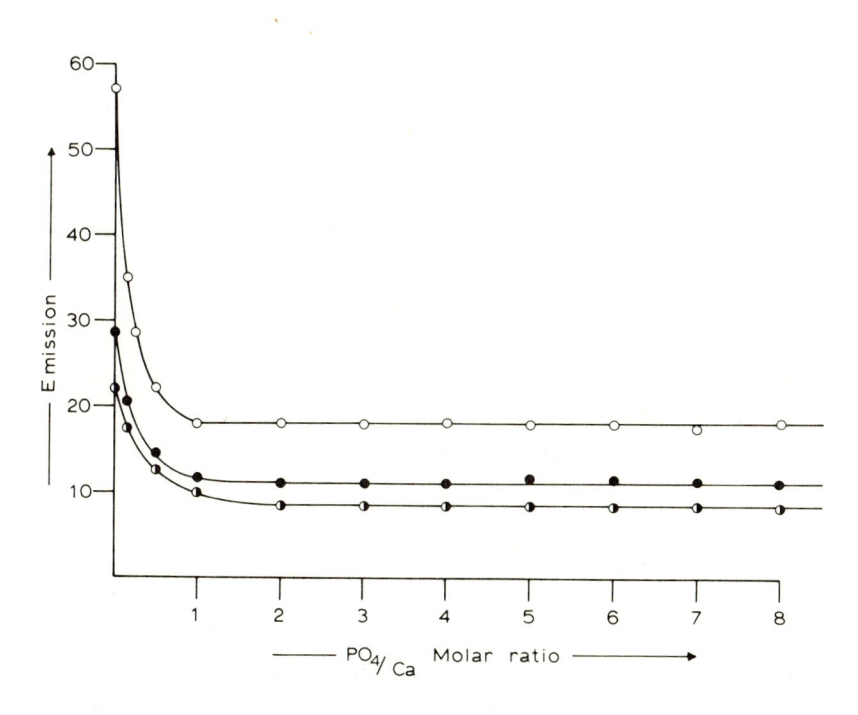

FIG. 21. INFLUENCE OF PHOSPHATE ON THE EMISSION OF CALCIUM. WAVELENGTH, 554 nm. ◐, 30 ppm Ca^{2+} + VARYING AMOUNTS OF PO_4^{3-}. ●, 40 ppm Ca^{2+} + VARYING AMOUNTS OF PO_4^{3-}. O, 80 ppm OF Ca^{2+} + VARYING AMOUNTS OF PO_4^{3-}. FROM REF. [56] WITH PERMISSION.

molar ratio of phosphate to calcium in solution but which is
dependent on the concentration of calcium. Further, the break
in the curve occurs at a phosphate/calcium molar ratio of one
in each case. These and other observations on the effect of
phosphate on calcium emission intensity have been discussed in
detail by Herrmann and Alkemade [57]. The atomic ratio of
phosphorous to calcium depends on several instrumental and
operational factors but it seems to be agreed that calcium
combines with phosphorous before entering the flame to yield
a solid particle of constant phosphorus/calcium ratio for a
given instrumental situation. This solid is of limited volatility in the flame and its rate of evaporation is invoked to
explain the observations. Fassel and Becker [58] have recently
shown that the interference can be completely eliminated or
reduced to a negligible level using techniques which will increase the rate of particle evaporation or increase the particle residence time in the flame as far as the detector is concerned. It would seem that we are left at the present time in
the unenviable position whereby flame photometric titrations
based on radiation interferences rest on using inefficient
atomizers and burners. Nevertheless, as we learn more about
ways to eliminate these interferences, we are also learning
more about means to optimize them for the purpose of titration.

B. Alkaline Earth-Phosphorus Systems

The first report of the use of flame photometry to detect
titration end points is the 1940 work of Torok [59]. A zinc
and hydrochloric acid chemical atomizer was used to generate
hydrogen gas which carried the sample into the flame of a
Bunsen burner. A spectroscope was used to observe the flame
emission. Figure 21 shows that the emission intensity upon
addition of phosphate to a calcium ion solution should decay

2. UNUSUAL END-POINT DETECTION METHODS

errors ranging from -5 to +1.6%. The method was turned around and phosphate was titrated with strontium to the appearance of the strontium line.

Erdey and Svehla [60] later used a Zeiss flame photometer, an acetylene-air flame, and a modified aspirator for end-point detection in the titration of calcium ion with phosphoric acid. This instrumentation produced a calcium to phosphate ratio of 1.5/1. Titrations curves were constructed by plotting the galvanometer reading (emission intensity) versus volume of titrant. A few tenths of a milligram of calcium ion could be titrated. Precision and accuracy were comparable to Torok's work.

C. The Lanthanum-Phosphate System

Svehla and Slevin [61] found that phosphate was an interference in the flame photometry of lanthanum and reported a flame photometric titration of lanthanum ion based on this fact and their finding that the lanthanum to phosphate ratio was 1/1. A Unicam SP 90 instrument was used with the following conditions reported: flame--acetylene-air; acetylene flow rate--900 ml/min; air flow rate--5 l/min; wavelength--560 nm; slit width--0.1 mm; burner height--2 cm; damping--lowest. The nature of flame photometric titrations results in some loss of sample but this can be kept to a minimum. Under the following conditions total sample consumption was kept to about 1 ml: not to zero but to some finite and constant level. Torok compensated for this by adjusting the slits of his spectroscope so it was only a matter of titrating until the signal corresponding to the alkaline earth metal emission could no longer be observed. One to five milliliter volumes of 1.0 N and 0.1 N solutions of barium, strontium, and calcium chlorides were titrated with ammonium dihydrogen phosphate with relative

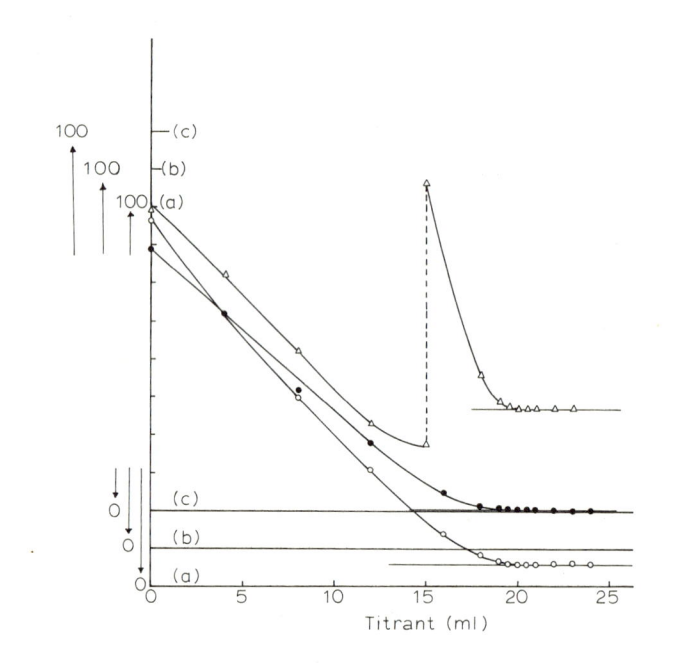

FIG. 22. FLAME PHOTOMETRIC TITRATION CURVES. O, 20.00 ml OF 0.1 M LaCl$_3$ TITRATED WITH 0.1 M (NH$_4$)$_2$HPO$_4$ (a). ●, 20.00 ml OF 0.01 M LaCl$_3$ TITRATED WITH 0.01 M (NH$_4$)$_2$HPO$_4$ (b). △, 20.00 ml OF 0.1 M LaCl$_3$ TITRATED WITH 0.1 M (NH$_4$)$_2$HPO$_4$, INCREASED SENSITIVITY BEING USED IN THE VICINITY OF THE END POINT (c). FROM REF. [61] WITH PERMISSION.

rate of sample consumption--4.5 ml/min; time required for galvanometer deflection--1.2 sec; region of data points--near the end point; number of galvanometer readings per titration--12 or less; initial sample volume--50 ml. The ratio of volume of sample consumed to total sample volume might indicate an error of -2% but the fact that the readings may be taken close to the end point greatly reduces the error. As a further aid

2. UNUSUAL END-POINT DETECTION METHODS   177

in locating the end point, the galvanometer sensitivity was increased in the region of the end point. Figure 22 shows titration curves obtained with and without increasing the sensitivity in this manner. For routine work it may be sufficient to locate the end point visually rather than plot the titration curves.

Acidic solutions of lanthanum chloride were titrated with diammonium hydrogen phosphate. A range of 2 to 0.5 mmoles of La(III) was titrated with an accuracy of -0.5 to -2%, respectively, and a precision (standard deviation) of about ±1 to 2%. A tenfold smaller range produced results of only slightly poorer precision and accuracy. Titrations of samples containing 0.02 mmoles of La(III) were accomplished with an accuracy of -5% and a precision of ±8%. Interferences investigated were: zirconium, titanium(III), cerium(III), yttrium(III), thorium, and aluminum. At concentrations nearly the same as that of lanthanum(III), the first two did not interfere, the second two change the slope of the titration curve after the end point but a graphical treatment could still be applied to determine the end point, and the last two were serious interferences. As might be expected, the method will not work in the presence of other rare earths but the absence of interference by either zirconium or titanium makes the method useful.

D. Conclusions

It seems to the writer that flame photometric titrations deserve more attention than they have received. The scope of the technique could be extended to most types of titration reactions if procedures can be devised to isolate the constituent producing the emission intensity before it is aspirated into the flame. Any phase separation process which removes a reactant or product from its product or reactant, respectively, would accomplish this goal. For example it should not be difficult to aspirate only the aqueous phase into the flame during a precipitation titration. Much is already known about the use of organic solvents in flame photometry and this knowledge could be used with extraction procedures for the phase separation required.

REFERENCES

1. D. J. Curran and J. L. Driscoll, Anal. Chem., 38, 1746 (1966).
2. J. L. Driscoll, Ph.D. Thesis, Dept. of Chem., Univ. of Mass., 1968.
3. J. R. Kronfeld, M.S. Thesis, Dept. of Chem., Univ. of Mass., 1966.
4. D. J. Curran and J. E. Curley, Anal. Chem., 42, 373 (1970).
5. K. S. Lion, Instrumentation in Scientific Research, McGraw-Hill, New York, 1959.
6. H. K. P. Nuebert, Instrument Transducers, Oxford University Press, London, 1963.
7. D. J. Curran, J. Chem. Educ., 46, A401, A465 (1959).
8. O. R. Gottlieb, Anal. Chim. Acta, 13, 101 (1955).
9. O. R. Gottlieb, Anal. Chem. Acta, 13, 531 (1955).
10. O. R. Gottlieb, Anal. Chim. Acta, 14, 497 (1956).
11. D. J. Curran and S. J. Swarin, Anal. Chem., 43, 358 (1971).
12. A. Goldstein, Proc. Amer. Acad. Arts Sci., 77, 237 (1949).
13. D. J. Curran and S. J. Swarin, Anal. Chem., 43, 1338 (1971).
14. O. R. Gottlieb, Anal. Chim. Acta, 13, 214 (1955).

15. J. F. Taylor and A. B. Hastings, J. Biol. Chem., 131, 649 (1931).
16. A. R. Blanchette, Ph.D. Thesis, Dept. of Chem., Univ. of Mass., 1971.
17. O. R. Gottlieb, Anal. Chim. Acta, 14, 24 (1956).
18. A. F. Krivis, E. S. Gazda, G. R. Supp, and P. Kipper, Anal. Chem., 35, 1955 (1963).
19. S. Bruckenstein and N. E. Vanderborgh, Anal. Chem., 38, 687 (1966).
20. N. E. Vanderborgh and W. D. Spall, Anal. Chem., 40, 256 (1968).
21. H. V. Malmstadt, C. G. Enke, and E. C. Toren, Jr., Electronics for Scientists, Benjamin, New York, 1962, pp. 354-355.
22. S. Bruckenstein and A. Saito, J. Amer. Chem. Soc., 87, 698 (1965).
23. S. Bruckenstein, Chem. Eng. News, 45, (29), 40 (1967).
24. C. D. Bogin, Ind. Eng. Chem., 16, 380 (1924).
25. T. Higuchi and K. A. Connors, "Phase Solubility Techniques," in Advances in Analytical Chemistry and Instrumentation, Vol. 4 (C. N. Reilley, Ed.), Interscience, New York, 1965, pp. 117-212.
26. A. Findlay and A. N. Campbell, The Phase Rule and its Applications, 8th ed., Dover, New York, 1945, pp. 212-216.
27. D. W. Rogers, Talanta, 9, 733 (1962).
28. D. W. Rogers and A. Ozsogomonyan, Talanta, 10, 633 (1963).
29. G. R. Atwood, in Encyclopedia of Industrial Chemical Analysis, Vol. 3 (F. D. Snell and C. L. Hilton, Eds.), Wiley (Interscience), New York, 1966, pp. 125-127.
30. E. R. Caley and A. E. Habboush, Anal. Chem., 33, 1613 (1961).
31. S. I. Spiridonova, Zh. Prikl. Khim., 20, 635 (1947); Chem. Abstr., 43, 6946c (1949).

32. S. I. Spiridonova, Zh. Prikl. Khim., 21, 948 (1948); Chem. Abstr., 44, 9216i (1950).
33. S. I. Spiridonova, Zh. Anal. Khim., 4, 169 (1949); Chem. Abstr., 44, 2886d (1950).
34. S. I. Spiridonova, Zh. Prikl. Khim., 25, 429 (1952); Chem. Abstr., 46, 6996c (1952).
35. S. I. Spiridonova, Zh. Fiz. Khim., 26, 1827 (1952); Chem. Abstr., 48, 12609c (1954).
36. S. I. Spiridonova and E. K. Nikitin, Izv. Vyssh. Ucheb. Zaved., Khim. Khim. Tekhnol., 1, 22 (1958); Chem. Abstr., 53, 9788c (1959).
37. S. I. Spiridonova, Trudy Saratovsk. Zoovet. Inst., 9, 399 (1959); Chem. Abstr., 56, 2938c (1962).
38. S. I. Spiridonova, Izv. Vyssh. Ucheb. Zaved., Khim. Khim. Tekhnol., 6, 343 (1963); Chem. Abstr., 59, 9674b (1963).
39. S. I. Spiridonova and E. K. Nikitin, Izv. Vyssh. Ucheb. Zaved., Khim. Khim. Tekhnol., 8, 31 (1965); Chem. Abstr., 63, 7629a (1965).
40. S. I. Spiridonova, Zh. Prikl. Khim., 32, 1268 (1959); Chem. Abstr., 53, 16826h (1959).
41. D. A. Dunnery and G. R. Atwood, Talanta, 15, 855 (1968).
42. D. W. Rogers, A. Ozsogomonyan, and A. Sumer, Talanta, 11, 507 (1964).
43. S. Siggia and J. G. Hanna, Anal. Chem., 21, 1086 (1949).
44. S. K. Suri, Talanta, 17, 577 (1970).
45. D. W. Rogers, D. Lillian, and I. D. Chawla, Mikrochim. Acta, 1968, 722.
46. D. W. Rogers, D. Lillian, and I. D. Chawla, Talanta, 13, 313 (1966).
47. D. W. Rogers, D. L. Thompson, and I. D. Chawla, Talanta, 13, 1389 (1966).
48. S. I. Spiridonova, Zh. Prikl. Khim., 19, 966 (1946); Chem. Abstr., 41, 5809h (1947).

2. UNUSUAL END-POINT DETECTION METHODS

49. S. I. Spiridonova, Zh. Prikl. Khim., 36, 2729 (1963); Chem. Abstr., 60, 12659g (1964).
50. D. W. Rogers and A. Ozsogomonyan, Talanta, 11, 652 (1964).
51. S. I. Spiridonova, Zh. Prikl. Khim., 22, 1284 (1949); Chem. Abstr., 45, 3281b (1951).
52. S. I. Spiridonova, Zh. Prikl. Khim., 25, 169 (1952); Chem. Abstr., 48, 8013b (1954).
53. S. I. Spiridonova, Zh. Fiz. Khim., 29, 159 (1955); Chem. Abstr., 50, 12726h (1956).
54. S. I. Spiridonova, Izv. Vyssh. Ucheb. Zaved., Khim. Khim. Tekhnol., 1, 51 (1958); Chem. Abstr., 52, 19674a (1958).
55. S. I. Spiridonova, Izv. Vyssh. Ucheb. Zaved., Khim. Khim. Tekhnol., 4, 186 (1961); Chem. Abstr., 55, 23165b (1961).
56. J. Yofe and R. Finkelstein, Anal. Chim. Acta, 19, 166 (1958).
57. R. Herrmann and C. T. J. Alkemade, Flame Photometry, 2nd ed., Wiley (Interscience), New York, 1963, pp. 300-307.
58. V. A. Fassel and D. A. Becker, Anal. Chem., 41, 1522 (1969).
59. T. Torok, Fresenius' Z. Anal. Chem., 119, 120 (1940).
60. L. Erdey and G. Svehla, Fresenius' Z. Anal. Chem., 154, 406 (1957).
61. G. Svehla and P. J. Slevin, Talanta, 15, 978 (1968).

AUTHOR INDEX

Numbers in brackets are reference numbers and indicate that an author's work is referred to although his name is not cited in the text. Underlined numbers give the page on which the complete reference is listed.

A

Ahluwalia, S.C., 46 [108], 84
Albert, A., 54 [141], 86
Alexander, W.A., 37 [85], 47 [114], 50 [85], 83, 85
Alkemade, C.T.J., 174 [57], 181
Alleman, T.G., 6 [11], 79
Allred, R.E., 75 [184], 89
Arenare, D., 52 [125], 58 [125], 60 [125], 62 [125], 85
Arnek, R., 47 [112], 85
Atwood, G.R., 140 [29], 143 [41], 145 [41], 146 [41], 165 [41], 179, 180

B

Baca, E.J., 54 [142], 86
Bark, L.S., 6 [31], 80
Bark, S.M., 6 [31], 80
Barthel, J., 61 [173, 174], 88
Bartholomew, C.H., 20 [44], 54 [44], 59 [44, 165], 81, 88
Basolo, F., 56 [146], 87
Bates, R.G., 56 [150, 151], 87
Becker, D.A., 174 [58], 181
Becker, F., 59 [164, 166], 61 [164, 173, 174], 88
Beezer, A.E., 5 [3], 6 [3], 79
Belisle, J., 40 [93], 84
Bell, J.M., 6 [9], 79
Ben-Yair, M.P., 29 [66], 82
Benzinger, T.H., 57 [162], 88
Bertrand, G.L., 61 [175], 88
Biggs, A.I., 54 [139], 86

Billingham, Jr., E.J., 33 [76], 36 [84], 44 [104, 105, 106], 50 [84], 83, 84
Blanchette, A.R., 118 [16], 120 [16], 179
Blank, C.L., 54 [142], 86
Bogin, C.D., 136 [24], 140 [24], 149 [24], 154 [24], 156 [24], 165 [24], 166 [24], 167 [24], 169 [24], 179
Bolles, T.F., 58 [172], 61 [172], 62 [172], 88
Branch, G.E.K., 54 [137], 86
Brandstetr, J., 29 [69], 82
Brenner, A., 47 [115], 48 [115], 116, 117], 61 [115], 85
Bricker, C.D., 6 [12], 22 [12], 23 [12], 29 [12], 37 [12], 79
Brown, H.C., 56 [147], 87
Brown, M.W., 23 [53], 29 [53], 82
Bruckenstein, S., 123 [19], 129 [19], 130 [19], 131 [19], 133 [19], 134 [22], 135 [23], 179
Brusin, S.S., 44 [103], 84
Bryson, A., 54 [138], 86
Burton, K.C., 21 [48], 34 [48], 81

C

Cabani, S., 61 [177], 89
Caley, E.R., 141 [30], 142 [30], 143 [30], 154 [30], 156 [30], 157 [30], 171 [30], 179
Campagnoli, J.M., 60 [169], 88
Campbell, A.N., 137 [26], 179

Carr, P.W., 6 [34], 24 [34, 54], 34 [81], 80, 82, 83
Chawla, I.D., 155 [45], 157 [46], 158 [46], 159 [45, 47], 160 [45], 165 [47], 180
Christensen, J.J., 3 [1], 5 [2], 6 [1], 10 [2, 35], 11 [35], 12 [36], 13 [2], 14 [2], 15 [39], 17 [1, 40, 42], 20 [35, 36, 43, 44, 46], 21 [46, 47], 27 [60], 28 [60], 33 [42, 78], 35 [82], 48 [118, 119, 120, 121], 50 [121, 123], 51 [121], 52 [123], 53 [119 121, 127], 54 [44, 47, 121, 128], 56 [43], 57 [156], 58 [123], 59 [44, 165], 60 [123, 168], 61 [121, 123, 156, 178], 63 [179, 180], 79, 81, 82, 83, 85, 86, 87, 88, 89
Clear, C.G., 54 [137], 86
Cobb, J.C., 40 [94], 84
Connors, K.A., 136 [25], 179
Conti, M., 54 [136], 86
Cope, B., 40 [91], 84
Cowell, C.F., 6 [9], 79
Crompton, T.R., 40 [91], 84
Cumming, A.C., 54 [130], 86
Curlcy, J.E., 94 [4], 112 [4], 121 [4], 178
Curran, D.J., 94 [1, 4], 97 [1], 98 [7], 101 [11], 102 [11], 103 [11], 104 [11], 105 [7], 106 [1], 107 [13], 108 [13], 109 [13], 110 [13], 112 [4], 113 [1], 114 [1], 118 [1, 11, 13], 121 [4], 122 [11, 13], 178

D

Daftary, R.D., 32 [73], 39 [73], 83

Datta, S.P., 56 [149, 152, 153], 87
Davidon, W.C., 52 [126], 85
Dean, J.A., 6 [19], 80
Dei, A., 58 [163], 88
DeLeo, A.B., 6 [16], 19 [16], 28 [62], 79, 82
Dobrokhotova, N.A., 44 [100], 84
Drago, R.S., 57 [159], 58 [159, 172], 61 [172], 62 [172], 87, 88
Driscoll, J.L., 94 [1, 2], 97 [1], 106 [1], 113 [1], 114 [1], 115 [2], 118 [1, 2], 178
Duer, W.C., 61 [175], 88
Dumbaugh, Jr., W.H., 28 [61], 82
Dunn, G.E., 54 [131], 86
Dunnery, D.A., 143 [41], 145 [41], 146 [41], 165 [41], 180

E

Eatough, D., 17 [42], 20 [44], 33 [42], 48 [118, 119, 120], 50 [123], 52 [123], 53 [119], 54 [44], 58 [123], 59 [44, 165], 60 [123], 61 [123, 171], 81, 85, 88
Enke, E.G., 130 [21], 179
Epley, T.D., 57 [159], 58 [159], 87
Erdey, L., 176 [60, 61], 181
Everson, W.L., 34 [79, 80], 39 [88, 89, 90], 40 [89, 90], 83, 84
Ewing, G.W., 6 [17, 29], 79, 80

F

Farquhar, E.L., 57 [158], 87
Fassel, V.A., 174 [58], 181
Felmeister, A., 29 [64], 82
Findlay, A., 137 [26], 179

AUTHOR INDEX

Finkelstein, R., 173 [56], 181
Forman, E.J., 41 [98], 84
Forsberg, J.H., 45 [107], 84
Freeberg, F.E., 29 [65], 82

G

Gaal, F.F., 44 [103], 84
Gallet, J.P., 29 [67, 68], 82
Gardner, W.L., 60 [170], 88
Gazda, E.S., 121 [18], 179
Gianni, P., 61 [177], 89
Gill, S.J., 57 [158], 87
Goldacre, R., 54 [141], 86
Goldstein, A., 103 [12], 178
Gottlieb, O.R., 98 [8], 99 [8], 100 [9, 10], 115 [14], 116 [8, 14], 117 [9, 10, 17], 118 [10], 120 [9], 121 [9, 17], 178, 179
Greathouse, L.H., 42 [99], 84
Grenthe, I., 47 [111], 85
Grundmann, R., 59 [166], 88
Grzybowski, A.K., 56 [149, 152, 153], 87

H

Habboush, A.E., 141 [30], 142 [30], 143 [30], 154 [30], 156 [30], 157 [30], 171 [30], 179
Haflinger, O., 56 [147], 87
Haldar, B.C., 32 [73], 39 [73], 83
Hale, J.D., 15 [39], 33 [78], 81, 83
Hanna, J.G., 150 [43], 166 [43], 167 [43], 170 [43], 180
Hansen, L.D., 10 [35], 11 [35], 13 [37], 15 [39], 17 [40], 20 [35, 45, 46], 21 [46, 47], 27 [45], 48 [121], 50 [121], 51 [121], 53 [45, 121, 127], 54 [47, 121, 142], 61 [121], 63 [179], 70 [182], 75 [184], 81, 85, 86, 88, 89
Harjanne, P., 54 [134], 86
Harmelin, M., 6 [32], 80
Harris, M.J., 25 [57], 82
Harris, P.C., 59 [167], 88
Hart, D.M., 47 [113], 85
Hastings, A.B., 115 [15], 179
Haydel, C.H., 42 [99], 84
Haymore, B.L., 60 [168], 88
Henry, R.H., 6 [25], 63 [25], 80
Herrmann, R., 174 [57], 181
Hetzer, H.B., 56 [151], 87
Higuchi, T., 136 [25], 179
Hume, D.N., 5 [4, 7], 6 [4, 13], 23 [13], 41 [98], 42 [13], 74 [4], 79, 84
Huyskens, P., 57 [160], 87

I

Irving, H.M.N.H., 21 [48], 34 [48], 81
Issa, K., 23 [53], 29 [53], 82
Izatt, R.M., 3 [1], 5 [2], 6 [1], 10 [2, 35], 11 [35], 12 [36], 13 [2], 14 [2], 15 [39], 17 [1, 40, 42], 20 [35, 36, 43, 44, 46], 21 [46, 47], 27 [60], 28 [60], 33 [42, 78], 35 [82], 48 [118, 119, 120, 121], 50 [121, 123], 51 [121], 52 [123], 53 [119, 121, 127], 54 [44, 47, 121, 128], 56 [43], 57 [156], 58 [123], 59 [44, 165], 60 [123, 168], 61 [121, 123, 156, 178], 63 [179, 180], 79, 81, 82, 83, 85, 86, 87, 88, 89

J

Jambon, C., 44 [102], 84
Janssen, H.J., 42 [99], 84
Javick, R.A., 29 [63], 82

Jespersen, N. D., 33 [75], 83
Johnston, H.D., 35 [82], 63 [180], 83, 89
Johnston, J., 54 [140], 86
Jordan, J., 5 [8], 6 [11, 18, 20, 22, 25, 27, 28, 29], 24 [54], 25 [56], 28 [61], 29 [63, 66], 33 [75, 76], 44 [104, 105, 106], 63 [25], 79, 80, 82, 83, 84

K

Keily, H.J., 6 [13], 23 [13], 42 [13], 79
Kenny, D., 70 [182], 89
Kettrup, A., 46 [110], 85
Kilpi, S., 54 [134], 86
King, E.J., 56 [154], 87
Kipper, P., 121 [18], 179
Kirby, A.H.M., 56 [148], 87
Kiss, T., 21 [49, 50], 22 [49, 51], 34 [49, 50], 44 [49], 81
Klockow, D., 21 [50], 34 [50], 81
Kronfeld, J.R., 94 [3], 178
Krivis, A.F., 121 [18], 179

L

Lamberts, L., 57 [160], 87
Lange, G., 61 [173], 88
Leden , I., 47 [111], 85
Leggate, P., 54 [131], 86
Lewis, E.A., 20 [45], 27 [45], 53 [45], 70 [182], 75 [184], 81, 89
Lillian, D., 155 [45], 157 [46], 158 [46], 159 [45], 160 [45], 180
Linde, H.W., 5 [4], 6 [4], 74 [4], 79
Lingane, J.J., 74 [183], 89
Lion, K.S., 98 [5], 178
Liquori, A.M., 54 [133], 86
Litchman, W.M. 70 [182], 75 [185], 89

Lumme, P.O., 54 [132], 86
Luschow, H.M., 59 [164], 61 [164, 173, 174], 88

M

Malmstadt, H.V., 130 [21], 179
Malspeis, L., 57 [157], 87
Martell, A.E., 17 [41], 81
Mash, C.J., 37 [85], 47 [114], 50 [85], 83, 85
Mattews, R.W., 54 [138], 86
McAuley, A., 37 [85], 47 [114], 50 [85], 83, 85
McCoy, R.D., 56 [145], 86
McCurdy, Jr., W.H., 6 [12], 22 [12], 23 [12], 29 [12], 37 [12], 79
McDaniel, D.H., 56 [147], 87
Mead, T.E., 41 [97], 84
Meier, J., 44 [104, 105, 106], 84
Merlin, J.C., 44 [102], 84
Merritt, L.L., 6 [19], 80
Metzger, Jr., W.H., 48 [116, 117], 85
Miller, F.J., 29 [70], 32 [71], 35 [83], 82, 83
Moeller, T., 45 [107], 84
Moore, T.E., 59 [167], 88
Moskovskaya, I.F., 44 [100], 84
Muller, R.H., 6 [10], 79
Murman, R.K., 56 [146], 87
Murphy, C.B., 6 [23], 80

N

Neerinck, D., 57 [160], 87
Nelson, D.P., 60 [168], 88
Neuberger, A., 56 [148], 87
Nielsen, T., 33 [77], 83
Nikitin, E.K., 141 [36, 39], 180
Nuebert, H.K.P., 98 [6], 178

O

Olofsson, G., 57 [161], 87
Oscarson, J.O., 57 [156], 61

AUTHOR INDEX

Oscarson, J.O., (continued), [156], 87
Owen, B.B., 56 [155], 87
Ozsogomonoyan, A., 140 [28], 141 [28], 147 [42], 148 [42], 149 [42], 161 [28], 162 [42], 163 [42, 50], 164 [28, 42, 50], 165 [42], 166 [28, 50], 169 [50], 171 [28], 179, 180, 181

P

Pack, R.T., 33 [78], 83
Paoletti, P., 52 [125], 58 [125, 163], 60 [125], 61 [125], 85
Papenmeier, G.J., 60 [169], 88
Papoff, P., 6 [24], 54 [143], 61 [176], 80, 86, 89
Paris, R.A., 29 [67, 68], 32 [72], 82, 83
Parkash, R., 46 [108], 84
Parker, R.D., 40 [92], 84
Partridge, J.A., 48 [121], 50 [121], 51 [121], 53 [121], 54 [121], 61 [121], 85
Paul, R.C., 46 [108], 84
Pei, P.T.S., 25 [56], 29 [63], 82
Peltier, D., 54 [136], 86
Pendergrast, J., 44 [104, 105, 106], 84
Periale, J., 40 [94], 84
Phillips, J.P., 6 [26], 80
Pinching, G.D., 56 [150], 87
Press, R.E., 32 [74], 83
Priestley, P.T., 6 [15, 33], 19 [15], 20 [15], 25 [58, 59], 29 [15], 36 [15], 37 [15], 79, 80, 82

Q

Quilty, C.J., 41 [95], 84

Quist, A.S., 54 [144], 86

R

Raffa, R.J., 57 [157], 87
Ragland, J.L. 5 [5], 32 [5], 79
Ramirez, E.M., 34 [79], 39 [89, 90], 40 [89, 90], 83, 84
Rasmussen, J.L., 33 [77], 83
Reed, A.H., 36 [84], 50 [84], 83
Reynolds, C.A., 25 [57], 82
Ripamonti, A., 54 [133], 86
Robinson, R.A., 54 [139], 86
Rogers, D.W., 138 [27], 139 [27], 140 [27, 28], 141 [28], 147 [42], 148 [42], 149 [42], 154 [27], 155 [45], 157 [27, 46], 158 [46], 159 [45, 47], 160 [45], 161 [28], 162 [28, 42], 163 [42, 50], 164 [28, 42, 50], 165 [42, 47], 166 [27, 28, 50], 169 [50], 171 [28], 179, 180, 181
Rogers, L.B., 5 [4], 6 [4], 74 [4], 79
Ruckman, J., 48 [118], 85
Rytting, J.H., 21 [47], 53 [127], 54 [47], 60 [168], 61 [178], 81, 86, 88, 89

S

Saito, A., 134 [22], 179
Salmon, H.I., 48 [116], 85
Sandhu, S.S., 46 [108], 84
Schafer, E., 39 [86], 83
Schmahl, N.G., 61 [173, 174], 88
Schmulbach, C.D., 47 [113], 85
Sebborn, W.S., 6 [33], 25 [58], 80, 82
Segatto, P.R., 6 [14], 19 [14], 46 [109], 79, 84

Selman, R.F.W., 6 [33], 25 [58], 80, 82
Sen, B., 13 [38], 28 [38], 81
Siggia, S., 150 [43], 166 [43], 167 [43], 170 [43], 180
Sillen, L.G., 17 [41], 52 [124], 81, 85
Sinclair, A.G., 23 [53], 29 [53], 82
Slevin, P.J., 175 [61], 176 [60, 61], 181
Snow, R.L., 50 [123], 52 [123], 58 [123], 60 [123], 61 [123], 85
Snyder, K.L., 41 [96], 84
Spall, W.D., 130 [20], 179
Specker, H., 46 [110], 85
Spiridonova, S.I., 141 [31, 32, 33, 34, 35, 36, 37, 38, 39], 142 [40], 143 [40], 160 [31, 32, 33, 35, 40, 48], 161 [49], 167 [34, 51], 168 [34], 169 [37, 51, 52, 53, 54], 170 [55], 179, 180, 181
Stern, M.J., 6 [16], 19 [16], 28 [62], 57 [157], 79, 82, 87
Stroh, H.H., 54 [129], 86
Sumer, A., 147 [42], 148 [42], 149 [42], 162 [42], 163 [42], 164 [42], 165 [42], 180
Supp, G.R., 121 [18], 179
Suri, S.K., 150 [44], 151 [44], 166 [44], 180
Svehla, G., 175 [61], 176 [60, 61], 176 [60, 61], 181
Swarin, S.J., 101 [11], 102 [11], 103 [11], 104 [11], 107 [13], 108 [13], 109 [13], 110 [13], 118 [11, 13], 122 [11, 13], 178
Swinehart, D.F., 56 [145], 86
Swithenback, J.J., 5 [6], 22 [6, 52], 43 [52], 44 [6], 79, 81

T

Takevchi, T., 39 [87], 83
Taylor, J.F., 115 [15], 179
Thomason, P.F., 29 [70], 32 [71], 35 [83], 82, 83
Thompson, D.L., 159 [47], 165 [47], 180
Tolman, D.O., 54 [128], 86
Topchieva, K.V., 44 [100], 84
Toren, Jr., E.C., 130 [21], 179
Torok, T., 174 [59], 181
Torsi, G., 54 [143], 61 [176], 86, 88
Tyrrell, H.J.V., 5 [3], 6 [3], 24 [55], 79, 82
Tyson, Jr., B.C., 6 [12], 22 [12], 23 [12], 29 [12], 37 [12], 79

V

Vacca, A., 52 [125], 58 [125, 163], 60 [125], 61 [125], 85
Vajgand, V.J., 44 [103], 84
Valcha, J., 44 [101], 84
Van Audenhaege, A., 57 [160], 87
Vanderborgh, N.E., 123 [19], 129 [19], 130 [19, 20], 131 [19], 133 [19], 179
Vanderzee, C.E., 54 [144], 86
Vaughn, G.A., 5 [6], 22 [6, 52], 43 [52], 44 [6], 79, 81
Vial, J., 32 [72], 83
Vlismas, T., 40 [92], 84

W

Wanders, A.C.M., 50 [122], 52 [122], 61 [122], 62 [122], 85
Wasilewski, J.C., 25 [56], 82
Watt, G.D., 33 [78], 83
Weiner, N.D., 29 [64], 82
Weisz, H., 21 [50], 22 [51],

AUTHOR INDEX

Weisz, H., (continued), 34 [50], 81
Wendlandt, W.W., 6 [21], 80
West, B.D., 54 [142], 86
Weston, B.A., 56 [152], 87
Westphal, G., 54 [129], 86
Wilde, E., 39 [86], 83
Willard, H.H., 6 [19], 80
Winkelblech, K., 54 [135], 86
Wrathall, D.P., 20 [43, 46], 27 [60], 28 [60], 54 [128], 56 [43], 57 [156], 60 [170], 61 [156], 81, 82, 86, 87, 88
Wu, W.C., 13 [38], 28 [38], 81

Y

Yamazaki, M., 39 [87], 83
Yofe, J., 173 [56], 181

Z

Zambonin, P.G., 5 [8], 6 [24], 54 [143], 61 [176], 79, 80, 86, 88
Zenchelsky, S.T., 6 [14, 30], 19 [14], 40 [94], 46 [109], 79, 80, 84
Zwietering, T.N., 50 [122], 52 [122], 61 [122], 62 [122], 85

SUBJECT INDEX

A

Acetamide, thermometric titration of, 42, 43
Acetanilide, thermometric titration of, 42
Acetate, by Continuous Flow Enthalpimetry, 27
Acetic acid
 in binary phase titrations, 158, 162, 163, 164
 by Continuous Flow Enthalpimetry, 26
 cryoscopic titration of, 131
 in ternary phase titrations, 167
 by thermometric titration, 28, 41
Acetic acid, glacial, as solvent in thermometric titrations, 41-43
Acetic anhydride, thermometric titration of, 42
Acetoacetic ester, in binary phase titrations, 166
Acetone
 in binary phase titrations, 157, 158, 160, 161, 162, 163, 164, 165, 166, 171
 calorimetric titration of, 58
 in ternary phase titrations, 166
Acetonitrile,
 calorimetric titration of, 58
 as solvent in thermometric titrations, 41, 42, 45, 46
Adenosine, calorimetric titration of, 53, 54
Adenosine diphosphate, thermometric titration of, 12, 13

Adenosine 5'-monophosphate, calorimetric titration of, 54
Adenosine triphosphate, calorimetric titration of, 57
Adipic dihydrazide, pressuremetric titration of, 122
Aldehyde C-8, in binary phase titrations, 159
Aldehyde C-12, in binary phase titrations, 159
Alkyl aluminum. See Organometallics
Alkyl amines
 cryoscopic titration of, 132, 133
 thermometric titration of 30, 32, 41, 42, 46
Allyl bromide, in binary phase titrations, 162, 163
Aluminum
 calorimetric titration of, 59
 thermometric titration of, 31, 32, 34
Amino acids, thermometric titration of, 28
m-Aminobenzoic acid, calorimetric titration of, 54
o-Aminobenzoic acid, calorimetric titration of, 54
p-Aminobenzoic acid, calorimetric titration of, 54
2-Amino-1-butanol, thermometric titration of, 30
m-Aminophenol, thermometric titration of, 32
Aminophylline, by thermometric titration, 28
Ammonia
 by Continuous Flow Enthalpimetry, 26
 thermometric titration of, 30
Ammonium hydroxide, thermometric titration of, 31

SUBJECT INDEX

Ammonium ion
  calorimetric titration of, 60
  pressuremetric titration of, 117, 118, 121, 122
  thermometric titration of, 31, 34
Aniline
  in binary phase titrations, 161, 164, 171
  thermometric titration of, 32
p-Anisidine, thermometric titration of, 42
Antimony pentachloride, calorimetric titration of, 57
Arsenite, pressuremetric titration of, 121
Aryl amines
  cryoscopic titration of, 132, 133
  thermometric titration of, 32, 38, 42
Azide, pressuremetric titration of, 117, 118, 121

B

Barium
  flame photometric titration of, 175-177
  thermometric titration of, 31
Barium chloride, by Continuous Flow Enthalpimetry, 26
Beckman thermometer, 6
Benzaldehyde, in ternary phase titrations, 166
Benzene
  in binary phase titrations, 156, 157, 159, 163, 164, 165
  as solvent in cryoscopic titrations, 132-134
  as solvent in thermometric titrations, 40, 41, 46
  in ternary phase titrations, 166, 167

Benzoic acid, thermometric titration of, 42
Benzyl amine, cryoscopic titration of, 132
Beryllium, thermometric titration of, 31
Binodal curve, 138-140, 150, 152
Bisulfite, thermometric titration of, 31, 37
Boric acid
  by Continuous Flow Enthalpimetry, 26
  by Direct Injection Enthalpimetry, 25
  thermometric titration of, 3-5, 28, 29
Bromide
  in molten salts, 44
  thermometric titration of, 34
Bromobenzene, in binary phase titrations, 159, 163
m-Bromobenzoic acid, thermometric titration of, 42
Bromoethane, in binary phase titrations, 159
Buret
  constant flow capillary, 74, 75
  motorized syringe, 74-76
n-Butanol
  in binary phase titrations, 156, 160, 165, 166
  in ternary phase titrations, 166
n-Butyl acetate, in binary phase titrations, 164, 165
n-Butyl bromide, in binary phase titrations, 161, 162, 164, 171
Butyllithium. See Organometallics

C

Cadmium
  calorimetric titration of, 59

SUBJECT INDEX

[Cadmium]
  thermometric titration of, 31
Calcium
  calorimetric titration of, 59
  flame photometric titration of, 173, 175-177
  thermometric titration of, 31, 33, 34
Calcium emission intensity, depression by phosphate, 173, 174
Calibration heater, 65, 66
Calorimeter vessel. *See* Reaction vessel
Carbohydrazide, pressuremetric titration of, 121, 122
Carbon dioxide, thermometric titration of, 43
Carbon disulfide, in binary phase titrations, 157
Carbon tetrachloride
  in binary phase titrations, 157, 159, 164, 166
  as solvent in thermometric titration, 40
  in ternary phase titrations, 167
Carboxylic acids, thermometric titration of, 42, 43
Carboxylic acid hydrazides, pressuremetric titration of, 119, 121
Catalysts, aprotonic acidity of, 44
Cerium(IV)
  pressuremetric titration of, 117
  thermometric titration of, 31, 34, 37
Cesium, calorimetric titration of, 60
Cetylpyridinium chloride, thermometric titration of, 29
CFE. *See* Continuous Flow Enthalpimetry

Chemical indicator
  furfural as, 142
  iodine as, 141
Chloride
  in molten salts, 44, 45
  thermometric titration of, 34, 42
Chloroacetic acid, by thermometric titration, 28
Chlorobenzene, in ternary phase titrations, 167, 170
m-Chlorobenzoic acid, thermometric titration of, 42
p-Chlorobenzoic acid, thermometric titration of, 42
1-Chlorobutane, in binary phase titrations, 159
Chloroform, in binary phase titrations, 159, 163, 164
Chlorpheniramine maleate, by thermometric titration, 28
Chloropromazine hydrochloride, by thermometric titration, 28
Chromate, in molten salts, 44
Chromium(III), thermometric titration of, 31
Citronella oil, in binary phase titrations, 159
Cobalt
  calorimetric titration of, 59, 60
  thermometric titration of, 31
Continuous Flow Enthalpimetry, 25-28
Continuous thermometric titration
  definition, 5
  uses of, 5, 6
Copper
  by Continuous Flow Enthalpimetry, 26
  thermometric titration of, 31, 32, 36
Copper(II)-phenanthroline, calorimetric titration of, 61

Copper(II)-pyridine, calorimetric titration of, 58
Cresol, thermometric titration of, 32
Cyanide
 pressuremetric titration of, 118
 thermometric titration of, 34
Cyclohexane
 in binary phase titrations, 159, 162, 165
 in ternary phase titrations, 166
Cyclohexylamine, cryoscopic titration of, 132
p-Cymene, in binary phase titrations, 157

D

Detergents, by thermometric titration, 29
Dibenzylamine, cryoscopic titration of, 132, 133
1,2-Dibromoethane, in binary phase titrations, 164
Di-n-butylamine, thermometric titration of, 42
Dichloromethane, in binary phase titrations, 162
DIE. See Direct Injection Enthalpimetry
Diethylamine, thermometric titration of, 42
N,N-diethylaniline, thermometric titration of, 42
Diethyl zinc. See Organometallics
Differential thermometric titration, 23
Di-isopropylamine, thermometric titration of, 42
β-Diketones, thermometric titration of, 43
N,N-Dimethylacetamide, calorimetric titration of, 58
N,N-Dimethylaniline, thermometric titration of, 42

N,N-dimethylbenzylamine, cryoscopic titration of, 132, 133
Dioxane
 in binary phase titrations, 157, 158, 161, 162, 163, 164, 166, 171
 in ternary phase titrations, 166
 thermometric titration of, 40
Direct Injection Enthalpimetry, 25-28
Di-sec-butylamine, thermometric titration of, 42
Dodecylamine, cryoscopic titration of, 132, 133
2-Dodecylbenzenesulfonate, by thermometric titration, 29

E

Electrolytes, phase titrations of, 167, 168
Enthalpy change, 3, 7, 8, 9, 25-28
Ephedrinium ion, calorimetric titration of, 57
Ethanol
 in binary phase titrations, 156, 157, 158, 159, 160, 161, 162, 163, 164, 165
 in ternary phase titrations, 166, 167
Ethanolamine
 in ternary phase titrations, 167
 thermometric titration of, 30
Ethers, thermometric titration of, 46
Ethyl acetate
 in binary phase titrations, 161
 calorimetric titration of, 58
Ethylenediamine, thermometric titration of, 30

SUBJECT INDEX

Ethyl ether, in ternary phase titrations, 167
Ethyl vinyl ether, in ternary phase titrations, 167
Equilibrium calculations, 24
Equilibrium constant, 7, 8, 23, 48-62

F

Faraday's law, 94, 96
Ferrocyanide, thermometric titration of, 34, 39
Fluoride, thermometric titration of, 34
Fructose, calorimetric titration of, 54

G

Gadolinium(III), thermometric titration of, 45, 46
Gallium, calorimetric titration of, 59
Gasoline, in binary phase titrations, 156
Geraniol Coeur, in binary phase titrations, 159
Geranylacetate, in binary phase titrations, 159
Gilmont micrometer syringe, 65, 75
Glycerol, in ternary phase titrations, 167
Glycinate, thermometric titration of, 13, 14
Glycine
  calorimetric titration of, 55, 56
  by Continuous Flow Enthalpimetri, 27
Glycine hydrochloride, 13-15
Grignard reagents. See Organometallics

H

Heat Capacity, 7, 9

Heat of reaction. See Enthalpy change
Henry's law, 95, 97
Hexylamine, cryoscopic titration of, 132, 133
Hydrazine
  as pressuremetric titrant, 113-115
  pressuremetric titration of, 117, 118, 120, 121
Hydrochloric acid
  by Continuous Flow Enthalpimetry, 26
  cryoscopic titration of, 131
  by Direct Injection Enthalpimetry, 25
  pressuremetric titration of, 116
  thermometric titration of, 3-5, 32
Hydrogen peroxide, pressuremetric titration of, 118
Hydrogen sulfate ion, calorimetric titration of, 54
Hydroquinone, thermometric titration of, 37, 38

I

Ideal gas law, 95, 96
Imidazole, calorimetric titration of, 55, 56
Imides, thermometric titration of, 43
Incremental thermometric titration
  definition of, 5
  uses of, 5, 6
Indium, calorimetric titration of, 59
Iodate
  pressuremetric titration of, 113-115, 116, 118
  thermometric titration of, 31
Iodide
  in molten salts, 44
  pressuremetric titration of, 117

SUBJECT INDEX

[Iodide]
  thermometric titration of, 31, 34, 38
Iodine
  pressuremetric titration of, 116
  thermometric titration of, 37
Ion aggregate formation, 134, 135
Ionization constant, 27, 28
α-Ionone, in binary phase titrations, 159
β-Ionone, in binary phase titrations, 160
Iron(II)
  calorimetric titration of, 59
  pressuremetric titration of, 117, 118, 120
  thermometric titration of, 31, 32, 35, 36, 39
Iron(III), thermometric titration of, 31, 32
Isoamyl alcohol, in binary phase titrations, 160, 161
Isonicotinic acid hydrazide, pressuremetric titration of, 121
Isopropanol, in binary phase titrations, 156, 157, 158, 161, 162, 163, 164, 171
Isothermal phase diagrams, 137

K

Kerosene, in binary phase titrations, 156

L

Lanthanum
  calorimetric titration of, 59
  flame photometric titration of, 175-177

Lead
  by Continuous Flow Enthalpimetry, 26
  by Direct Injection Enthalpimetry, 25
  thermometric titration of, 32
Lemongrass oil, in binary phase titrations, 159
Lithium, thermometric titration of, 31

M

Magnesium
  calorimetric titration of, 59
  by Direct Injection Enthalpimetry, 25
  thermometric titration of, 31, 33
Malonate, thermometric titration of, 10-12
Manganese
  calorimetric titration of, 59
  by Continuous Flow Enthalpimetry, 26
  thermometric titration of, 31
Manometer, 98
  "Electronic manometer," 105-107
  free type, 101-105
  horizontal column type, 100-101
  U-tube type, 98-100
Mercury(II), thermometric titration of, 31, 33, 34
Mercury(II)-di-2-aminoethanol, calorimetric titration of, 61
Mercury(II) perchlorate
  thermometric titration with chloride ion, 15-17
  thermometric titration with cyanide ion, 17, 18
Mesitylene, in binary phase titrations, 156, 157

SUBJECT INDEX

Metanilic acid, calorimetric titration of, 55, 56
Methanol
 in binary phase titrations, 156, 157, 158, 160, 161, 162, 163, 164, 165, 171
 in ternary phase titrations, 170
Methyl acetate, in binary phase titrations, 160
Methyl-*m*-aminobenzoic acid, calorimetric titration of, 54
Methyl-*o*-aminobenzoic acid, calorimetric titration of, 54
Methyl-*p*-aminobenzoic acid, calorimetric titration of, 54
2-Methyl-2-amino-1,3-propanediol, thermometric titration of, 30
2-Methyl-2-amino-1-propanol, thermometric titration of, 30
Methyl aniline, in binary phase titrations, 161, 164, 171
Methyl isobutyl ketone, in binary phase titrations, 165
Methyl salicylate, in binary phase titrations, 159
Metol, thermometric titration of, 38
Miscibility boundary, 138
Molar freezing point constant, 124
Molten salts, as solvents in thermometric titrations, 44, 45

N

Naphthalene, as solvent in cryoscopic titrations, 134

Nickel
 calorimetric titration of, 59, 60
 by Continuous Flow Enthalpimetry, 26
 thermometric titration of, 31, 33
Nicotinamide, by thermometric titration, 28
Nicotine, thermometric titration of, 32
Nitric acid, by Continuous Flow Enthalpimetry, 26
Nitrobenzene, in binary phase titrations, 161, 164, 171
m-Nitrobenzoic acid, thermometric titration of, 42
Nitroethane, in binary phase titrations, 157, 158
Nitromethane
 in binary phase titrations, 158, 159
 in ternary phase titrations, 166
1-Nitropropane, in binary phase titrations, 158
2-Nitropropane, in binary phase titrations, 158, 159

O

Organometallics
 alkyl aluminum, 38, 40
 butyllithium, 38
 diethyl zinc, 38, 40
 Grignard reagents, 40
 thallous ethoxide, 40
 trimethyl tin chloride, 57
Orthophosphate, by Continuous Flow Enthalpimetry, 26
Ovalbumin, titration of prototropic groups on, 33
Oxyhemoglobin, pressuremetric titration of, 115, 118, 120

P

Palladium(II), thermometric titration of, 34
Periodate, pressuremetric titration of, 116
Petroleum ether, in binary phase titrations, 156
Petroleum products, total acidity in, 41
Pharmaceuticals, by thermometric titration, 28
Phase rule, 136
Phenidone, thermometric titration of, 38
Phenol
  calorimetric titration of, 57
  by Continuous Flow Enthalpimetry, 27
  thermometric titration of, 32, 43
Phenols, of coal, 44
Phenylacetic acid hydrazide, pressuremetric titration of, 121
Phenylethylacetate, in binary phase titrations, 159
Phenylethylether, in binary phase titrations, 159
Phosphate
  calorimetric titration of, 54
  thermometric titration of, 31
  as titrant in flame photometric titrations, 174-177
Piperidine, cryoscopic titration of, 132, 133
Pit mapping, 52
Plait point, 140, 141, 145
Potassium
  calorimetric titration of, 59, 60
  thermometric titration of, 34
Potassium hydroxide, by Continuous Flow Enthalpimetry, 26

Pressure transducer, 93, 96, 97, 98-113
  capacitive type, 106-107
  differential, 107-110
  Piezoelectric type, 107-110
n-Propanol, in binary phase titrations, 156, 157, 165
Pseudoephedrinium ion, calorimetric titration of, 57
Purine, calorimetric titration of, 57
Pyridine
  in binary phase titrations, 159, 166
  calorimetric titration of, 55, 56, 58
  by Continuous Flow Enthalpimetry, 26, 27
  in ternary phase titrations, 166, 167
  thermometric titration of, 30, 32, 42

R

Rare earth metals, calorimetric titration of, 59
Reaction vessel, 63, 65, 68, 72, 75, 77
Resorcinol, thermometric titration of, 38
Ribose, calorimetric titration of, 53, 54
Rubidium, calorimetric titration of, 60

S

Sage pump, 63
Scandium, calorimetric titration of, 59
Schematic mapping, 52
Silver, thermometric titration of, 30, 31, 32, 33, 34, 44
Silver nitrate, cryoscopic titration of, 131
Silver(I) pyridine, calorimetric titration of, 58-60

SUBJECT INDEX

Sodium bicarbonate, thermometric titration of, 31
Sodium carbonate
  pressuremetric titration of, 117, 166
  thermometric titration of, 31
Sodium dodecylsulfate, calorimetric titration of, 60
Sodium hydroxide
  by Continuous Flow Enthalpimetry, 26
  pressuremetric titration of, 116, 117, 121
  thermometric titration of, 30, 31, 41
Sodium ion, calorimetric titration of, 59
Sodium tetraborate, thermometric titration of, 31
Solubility titrations, 136
Strontium
  flame photometric titration of, 175-177
  thermometric titration of, 31
Sulfamic acid, pressuremetric titration of, 117
Sulfide, thermometric titration of, 34
Sulfite
  by Continuous Flow Enthalpimetry, 26
  thermometric titration of, 37
Sulfonic acid amides, thermometric titration of, 38
Sulfuric acid, pressuremetric titration of, 116

T

Thallium(I), thermometric titration of, 34
Thallous ethoxide. *See* Organometallics
Theophylline, by thermometric titration, 28
Thermistor, 6, 25, 63, 65, 66, 68, 70, 126, 128-130
Thermochemical indicator, 20-22
  acetone as, 22, 43, 44
  Arsenic(III) as, 21
  cerium(IV) as, 21
  glucose as, 21
  hydrogen peroxide as, 21, 22, 44
  resorcinol as, 44
  sulfate ion as, 20, 22
Thiocyanate, thermometric titration of, 34
Thiosulfate, thermometric titration of, 31, 38
Thiourea, thermometric titration of, 37
Thorium(IV), thermometric titration of, 32
Tie line, 140, 141
Tin(II), pressuremetric titration of, 117
Tin(IV), thermometric titration of, 31, 46
Titanium(III), pressuremetric titration of, 118, 120
Toluene, in binary phase titrations, 156, 159, 163
p-Toluic acid hydrazide, pressuremetric titration of, 121
o-Toluidine, thermometric titration of, 32
p-Toluidine, thermometric titration of, 42
1,2,3-Triazole-4,5-dicarboxylic acid, calorimetric titration of, 54
Tri-n-butylamine, thermometric titration of, 42
Trichloroacetic acid, by thermometric titration, 28
Triethylamine, calorimetric titration of, 58
Trimethyl tin chloride. *See* Organometallics
Tris(hydroxymethyl)aminomethane calorimetric titration of, 55, 56

[Tris(hydroxymethyl)amino-
  methane]
  thermometric titration of,
    30
Turpentine, in binary phase
  titrations, 161

U

Urea
  calorimetric titration of,
    57
  thermometric titration of,
    42
Uranium(VI), thermometric
  titration of, 32, 35, 36

V

van't Hoff i factor, 124-128
Variable metric minimization,
  52

W

Water
  in binary phase titrations,
    160, 165, 166
  by Direct Injection
    Enthalpimetry, 25
  in ternary phase titra-
    tions, 166, 167, 170

[Water]
  thermometric titration of,
    42
Wheatstone bridge, 63, 65, 71,
  74, 126, 128-130

X

Xenon trioxide, as pressuremetric titrant, 120
Xylene, in binary phase titrations, 156, 157, 159, 163

Y

Yttrium, calorimetric titration of, 59

Z

Zinc
  calorimetric titration of,
    59
  thermometric titration of,
    31, 32
Zinc(II)-phenanthroline,
  calorimetric titration of,
    61
Zirconium(IV), thermometric
  titration of 32